The 2016 Mw 7.1 Kumamoto Earthquake

Aiming Lin

The 2016 Mw 7.1 Kumamoto Earthquake

A Photographic Atlas of Coseismic Surface Ruptures Related to the Aso Volcano, Japan

Springer

Aiming Lin
Department of Geophysics
Kyoto University
Kyoto, Japan

Based on a translation from the Japanese language edition:
2016nen M7.3 Kumamoto Jishin: Chihyo Jishin Danso to Aso Kazan niokeru Visual Kiroku by Aiming Lin

ISBN 978-981-10-5854-7 ISBN 978-981-10-5855-4 (eBook)
https://doi.org/10.1007/978-981-10-5855-4

Library of Congress Control Number: 2017952049

Printed on acid-free paper

This Springer imprint is published by Springer Nature
The registered company is Springer Nature Singapore Pte Ltd.
The registered company address is: 152 Beach Road, #21-01/04 Gateway East, Singapore 189721, Singapore

Preface

The M_w 7.1 (Mj 7.3) Kumamoto earthquake occurred on April 16, 2016, resulting in more than 100 deaths and extensive damage on Kyushu Island in southwestern Japan. In order to determine the motion of the seismogenic fault, ground deformation, and relationships between coseismic surface ruptures and pre-existing faults, our survey group conducted fieldworks immediately, beginning one day after the main shock. During this time, we observed the principal structural features and measured offsets at the main locations of the coseismic surface ruptures, and then we retrieved ground deformation markers from locations that were damaged during the earthquake. Field investigations revealed that the coseismic surface rupturing terminated at the Aso caldera, and suggested that the newly formed coseismic ruptures under the Aso caldera are potential new channels for magma venting, which change the spatial heterogeneity and mechanical properties of the Aso volcano, and therefore may necessitate reassessment of the volcanic hazard in the vicinity of the Aso volcano (Lin et al. 2016). Amazingly, as suggested in our previous study (Lin et al. 2016), the Aso volcano re-erupted on October 8, 2016, after a 36-year dormant duration.

The preliminary fieldworks were reported in our previous publications (Lin et al. 2016; Lin 2017; Lin and Chiba 2017). Subsequently, to understand the seismotectonic history of the seismogenic fault zone—including the recurrence interval of morphogenic earthquakes, the slip rate, and the source fault characteristics of the 2016 Kumamoto earthquake—our group has also conducted paleoseismic studies, including fieldworks and trenching investigations on the 2016 source seismic faults and related active faults (Lin et al. 2017). This photographic atlas records the main deformation characteristics of the coseismic surface ruptures and the nature of the source seismogenic faults and related earthquake disasters, mainly based on the photographs taken during our field investigations and the related studies conducted in the year after the Kumamoto earthquake. This book is intended not only for geologists, seismologists, and engineers, as a means of furthering their understanding of the seismic mechanisms and surface rupture deformation characteristics of large intracontinental earthquakes, but also for advanced undergraduate and graduate students, as a textbook. A similar book, *The Great Wenchuan Earthquake of 2008: A Photographic Atlas of Surface Rupture and Related Disaster*, was published 1 year after the 2008 great Wenchuan earthquake (Lin and Ren 2009) and has been widely acclaimed by the earth science community. I hope this book will provide an important background for seismic hazard evaluation in the densely populated Kumamoto region.

I am grateful to the many organizations and individuals who helped to make this book possible within the short time of 1 year after the earthquake. I would particularly like to thank F. Nakajima, T. Satsukawa, M. Wang, Zahra Mohammadi Asl, D. Bian, R. Fueta, P. Chen, K. Sado, N. Takahashi, S. Hirata, K. Osono, and K. Fukuoka, for their assistance in the field. Thanks are also due to T. Chiba for providing the red relief maps and LiDAR images, K. Matsumoto and K. Tanoue for their kind permission to do trench investigations on their private lands, N. Akiyama and T. Sugita for arranging the trenching investigation, and Y. Yoshida for illustrating and editing figures.

Kyoto, Japan — Aiming Lin
May 2017

References

Lin A (2017) Structural features and seismotectonic implications of coseismic surface ruptures produced by the 2016 Mw 7.1 Kumamoto earthquake. J Seismol 21:1079–1100. https://doi.org/10.1007/ s10950-017-9653-5

Lin A, Chiba T (2017) Coseismic conjugate faulting structures associated with the 2016 Mw 7.1 Kumamoto earthquake, Japan. J Struct Geol 9:20–30. Open access article http://creativecommons.org/licenses/by-nc-nd/4.0/

Lin A, Ren Z (2009) The Great Wenchuan earthquake of 2008—a photographic Atlas of surface rupture and related disaster. Springer, Berlin, 121p. ISBN:978-3-642-03758-0

Lin A, Satsukawa T, Wang M, Mohammadi Asl Z, Fueta R, Nakajima F (2016) Coseismic rupturing stopped by Aso volcano during the 2016 Mw 7.1 Kumamoto earthquake, Japan. Science 354:869–875. https://doi.org/10.1126/science.aah4629

Lin A, Chen P, Satsukawa T, Sado K, Takahashi N, Hirata S (2017) Millennium recurrence interval of morphogenic earthquakes on the seismogenic fault zone that triggered the 2016 Mw 7.1 Kumamoto earthquake, SW Japan. Bull Seismol Soc Am, in press. doi:https://doi.org/10.1785/0120170149

Contents

List of Figures

Figure Plates

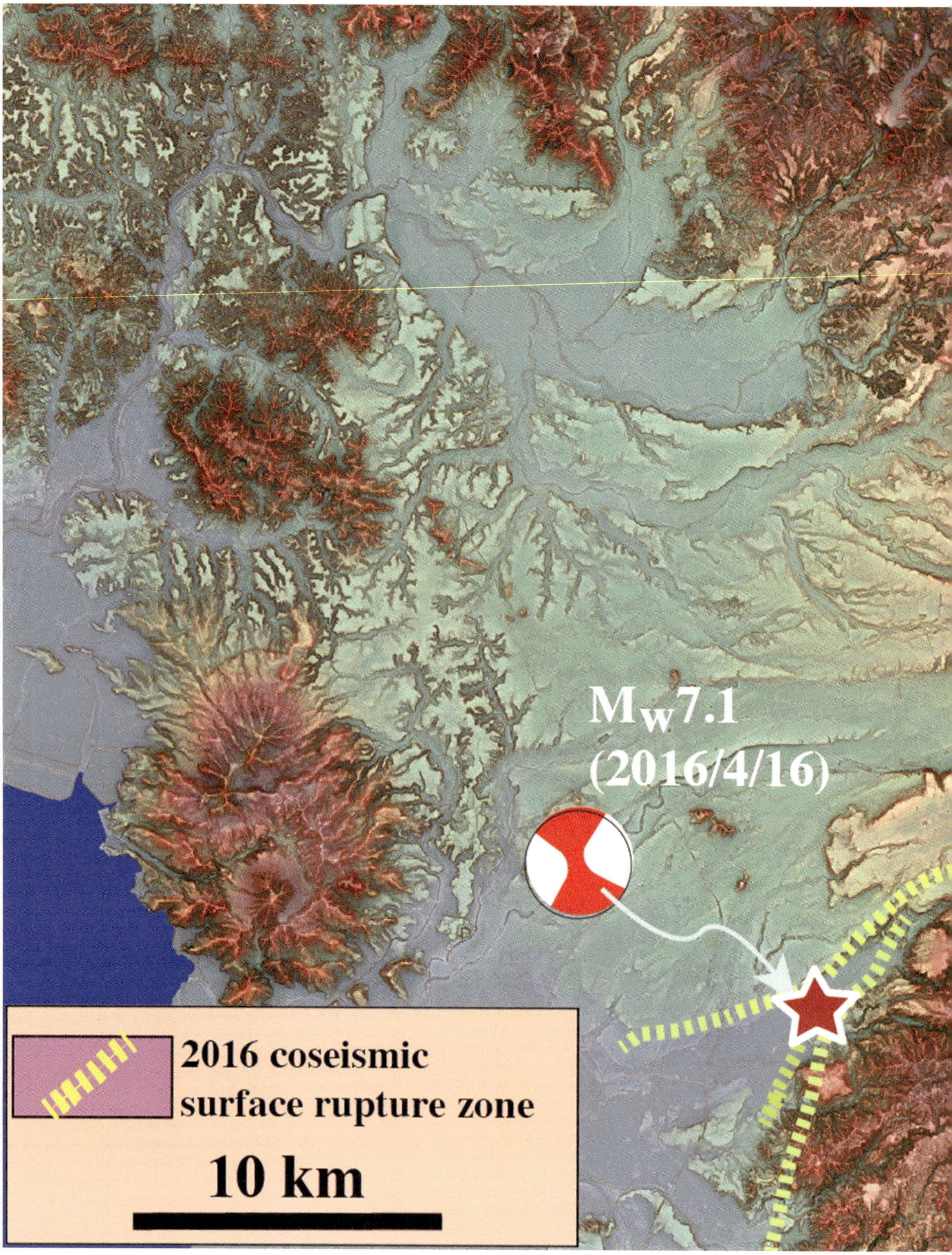

Figure Plate 1 Red relief image map of the study area around the Aso volcano clusters and 2016 Kumamoto coseismic surface rupture zone (image courtesy of T. Chiba)

N
Aso caldera

Figure Plate 2 Drone image showing a coseismic graben structure on the western side of the Aso caldera

a

Figure Plate 3 Right-lateral strike-slip offset of wheat fields in the coseismic surface rupture zone. (**a**) Aerial photograph (with permission from The Mainichi Newspapers). (**b**, **c**) Strike-slip offset of field paths, measuring ~2.5 m, which was the maximum displacement in the 2016 coseismic surface rupture zone

b

c

Figure Plate 3 (continued)

a

Figure Plate 4 Damaged building (**a**) and house (**b**) on the same coseismic fault scarp (vertical offset ~0.7 m)

b

Figure Plate 4 (continued)

a

Figure Plate 5 (**a**) Coseismic surface ruptures and a damaged building in the schoolyard at the Aso campus of Tokai University. (**b**) Mole track structure in the surface rupture zone in the schoolyard

b

Figure Plate 5 (continued)

Figure Plate 6 Drone image showing an overview of the Komezuka volcanic cone ruptured by the 2016 Kumamoto earthquake

a

b

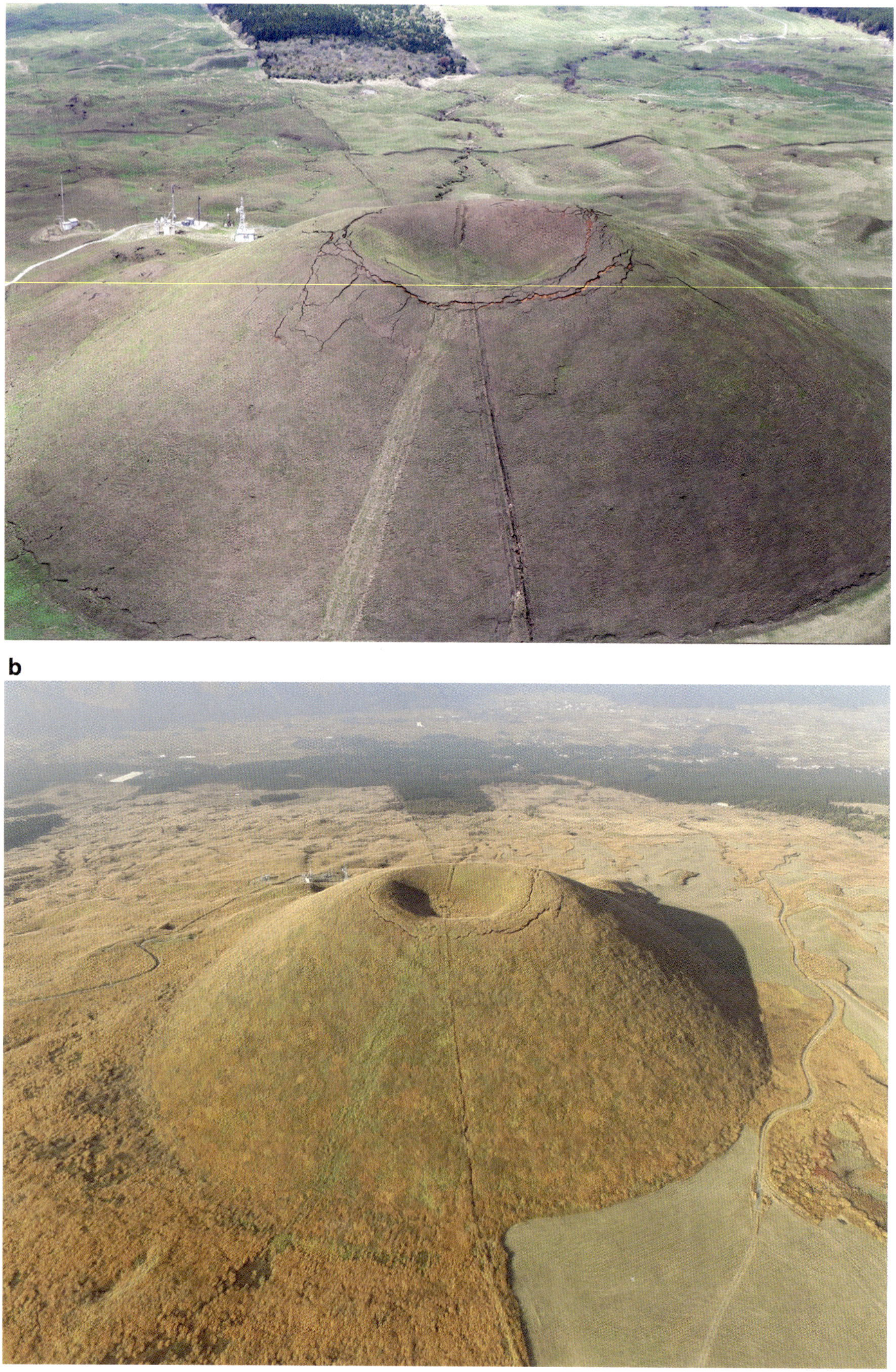

Figure Plate 7 Coseismic surface ruptures on both the top and the foot of the Komezuka caldera. (**a**) Aerial photograph acquired on April 17, 2016 (with kind permission from Jiji Press). (**b**) Drone image taken on November 16, 2016

a

b

Figure Plate 8 Drone images showing (**a**) the phreatic explosion of the Aso volcano after the 2016 Kumamoto earthquake, and (**b**) a close-up view of the explosion (taken on November 16, 2016)

a

Figure Plate 9 Stripped sections of exposure walls from the Takagi trench. The trench was dug across the coseismic surface rupture along the Hinagu Fault at the Takagi site, Mihune town. (**a**) North wall. (**b**) South wall

b

Figure Plate 9 (continued)

Figure Plate 10 Car struck by a large rock displaced during the Kumamoto earthquake

Figure Plate 11 Aso Shrine collapsed by the Kumamoto earthquake (**a, b**). (**a**) Overview of damaged Aso shrine. (**b**) Damaged shrine tower gate

An Overview of the 2016 M_w 7.1 Kumamoto Earthquake

1

The M_w 7.1 (Mj 7.3) Kumamoto earthquake occurred on 16 April 2016, resulting in extensive damage and more than 100 deaths on Kyushu Island, Japan. With a seismic rupture zone ≥40 km long and a magnitude (M_w) of 7.1, this shock was the largest inland earthquake recorded in Japan Islands in the past century. The main shock was accompanied by > 1000 foreshocks and aftershocks during the week of 14–20 April 2016 (Japan Meteorological Agency 2016). Three $M_w \geq 5.5$ foreshocks occurred 2 days before the main shock, including an M_w 6.2 (Mj 6.5) and M_w 5.5 (Mj 5.7) on 14 April 2016 and an M_w 6.0 (Mj 6.4) on 15 April 2016 (Fig. 1). Subsequently, four $M_w > 5.0$ aftershocks occurred within 6 h of the main shock on 16 April 2016 (Japan Meteorological Agency 2016a). Epicenters migrated from southwest to northeast, mostly along preexisting active faults and throughout Aso caldera in the first 3 days after the main shock (Fig. 1).

Seismic inversion results suggest that (i) the focal depth was ~15 km and (ii) the earthquake had a predominantly strike–slip focal mechanism on a fault striking NE–SW and dipping SE at ~80°, with a compression axis oriented E–W (National Research Institute for Earth Science and Disaster Prevention 2016; Yagi et al. 2016), However, surface deformation in the northeastern segment of the coseismic surface rupture zone was characterized by normal-dominated displacement that formed graben structures extending for ~10 km within the west–southwest side of Aso caldera.

Interferometric Synthetic Aperture Radar (InSAR) data revealed that the total rupture length of the seismogenic fault was ≥40 km with a maximum offset of up to >1.0 m (Geospatial Information Authority of Japan 2016) consistent with our field investigations (Fig. 2). The InSAR analysis shows that (i) the distinct ground deformation occurred in two sections, one along the southwest and SW–central segments that are dominated by strike–slip movement along the preexisting active faults of the Hinagu–Futagawa fault zone (HFFZ) and the other along the northeast segment of the coseismic surface rupture zone that is dominated by normal faults as graben structures inside Aso caldera (Fig. 2).

The geological evidence and seismic and InSAR data demonstrate that the distribution patterns and coseismic displacements as well structural features of coseismic surface ruptures are controlled by preexisting active Hinagu-Futagawa fault zone and active faults inside Aso caldera.

A. Lin, *The 2016 Mw 7.1 Kumamoto Earthquake*, https://doi.org/10.1007/978-981-10-5855-4_1

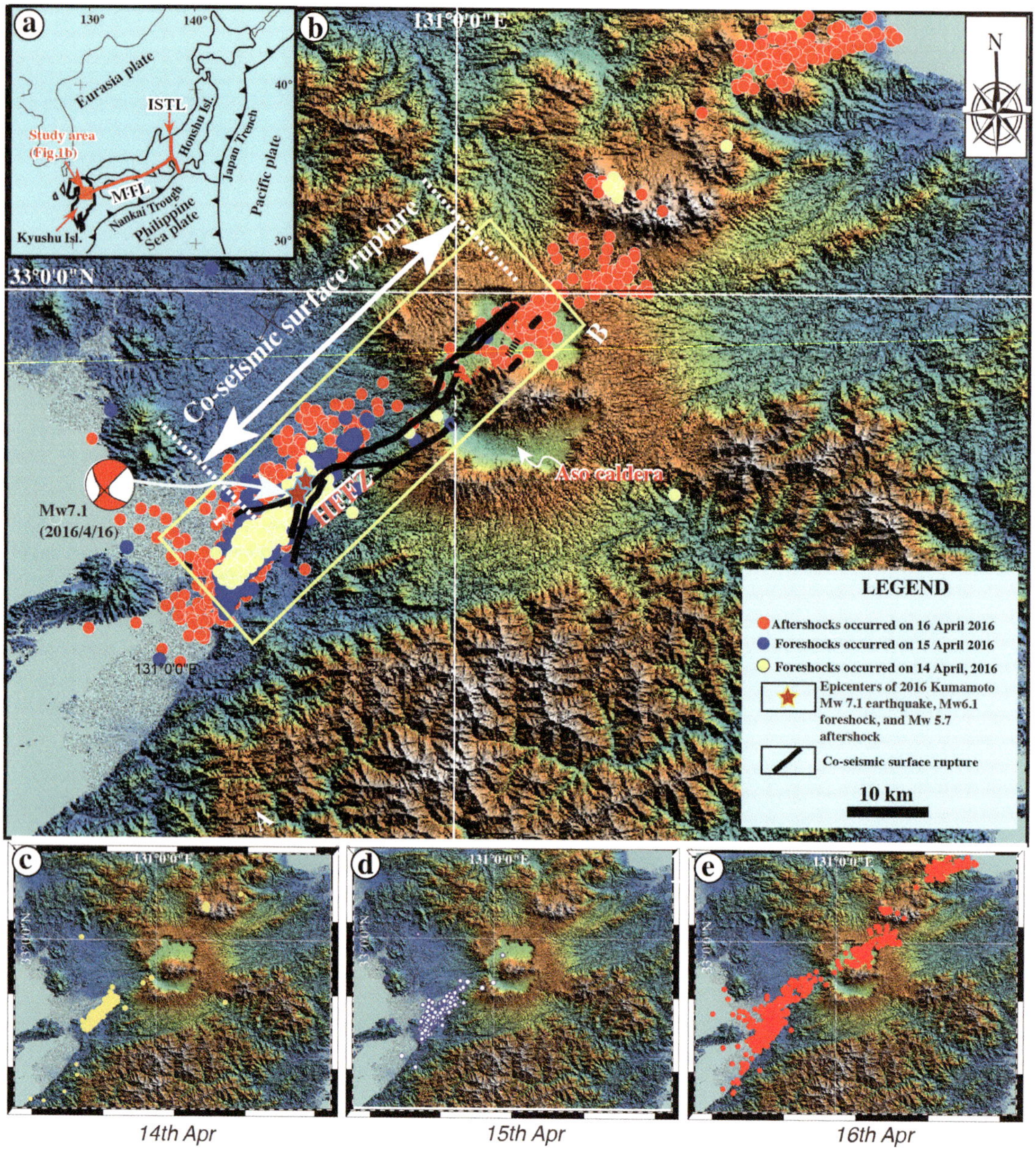

Fig. 1 Index map showing the distribution of the 2016 Kumamoto coseismic surface ruptures [modified from Lin et al. (2016) and Lin (2017)]. (**a**) Tectonic setting map. (**b**) *Color-shaded* relief map showing the distribution of the coseismic surface ruptures, foreshocks, and aftershocks that occurred in the period during 14 and 16 May 2016 and (**c**)-(**e**) the epicentral distribution of earthquakes that occurred on 14, 15, and 16 April 2016, respectively (epicentral data cited from the National Research Institute for Earth Science and Disaster Resilience, 2016). Profile A–B is shown in Fig. 4c. MTL, Median Tectonic Line; ISTL, Itoigawa-Shizuoka Tectonic Line; Honshu Isl., Honshu Island; Kyushu Isl., Kyushu Island; HFFZ, Hinagu-Futagawa fault zone

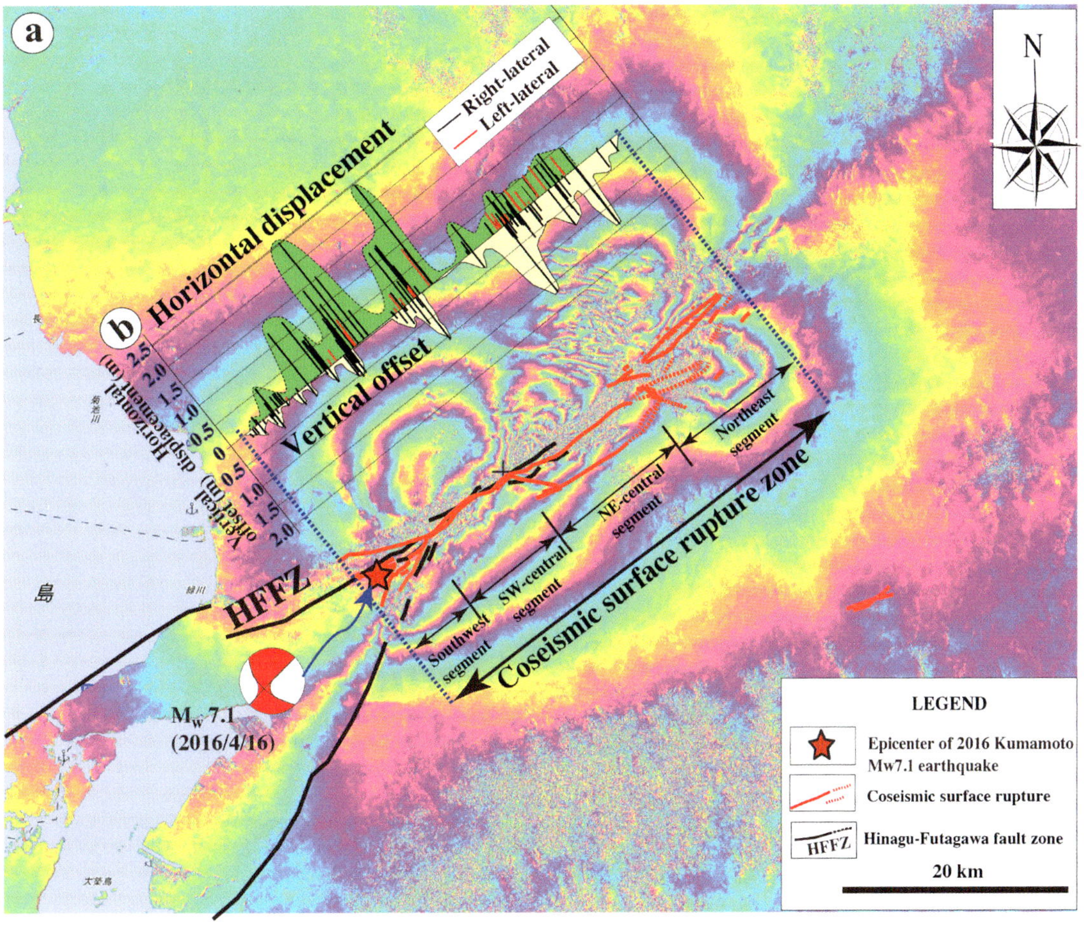

Fig. 2 InSAR image and distribution map showing the ground deformation features and distribution pattern of coseismic displacements (modified from Lin 2017). (**a**) InSAR image generated from PALSAR-2 data (acquired on 16 January 2016 and 20 April 2016) showing the deformation features of ground surface caused by the 2016 Kumamoto earthquake [modified from the Geospatial Information Authority of Japan (2016)]. The coseismic surface displacement zone is up to ~40 km in length. Active faults cited from the Geographical Survey Institute (2001) and Headquarters for Earthquake Research Promotion (2016). (**b**) The field-measured offset distribution along the coseismic surface rupture zone is shown for comparison. SW segment, southwest segment; NE segment, northeast segment

References

Geographical Survey Institute (2001) Active fault map in urban area, Kumamoto, 1:25,000

Geospatial Information Authority of Japan (2016) Information concerning with the 2016 Kumamoto earthquake. http://www.jishin.go.jp/regional_seismicity/rs_katsudanso/f093_futagawa_hinagu/

Headquarters for Earthquake Research Promotion (2016) Evaluation of Heisei 28 (2016) Kumamoto earthquake. http://www.mext.go.jp/component/a_menu/science/detail/__icsFiles/afieldfile/2016/04/15/1285728_001_1.pdf

Lin A, Satsukawa T, Wang M, Asl ZM, Fueta R, Nakajima F (2016) Coseismic rupturing stopped by Aso volcano during the 2016 Mw 7.1 Kumamoto earthquake, Japan. Science 354:869–875. https://doi.org/10.1126/science.aah4629

Lin A (2017) Structural features and seismotectonic implications of coseismic surface ruptures produced by the 2016 Mw 7.1 Kumamoto earthquake. J Seismol, i 21:1079–1100. https://doi.org/10.1007/s10950-017-9653-5

Japan Meteorological Agency (2016) Report on the Heisei 28 (2016) Kumamoto earthquake (No. 10). http://www.jma.go.jp/jma/menu/h28_kumamoto_jishin_menu.html

National Research Institute for Earth Science and Disaster Resilience (2016) 16 April 2016 Kumamoto Earthquake and its aftershock distribution. http://www.hinet.bosai.go.jp/topics/nw-kumamoto160416/?LANG=ja

Yagi Y, Okuwaki R, Enescu B, Kasahara A, Miyakawa A, Otsubo M (2016) Rupture process of the 2016 Kumamoto earthquake in relation to the thermal structure around Aso volcano. Earth Planets Space 68:118

2 Tectonic Setting

The study area is located in central Kyushu Island, around the west and central side of Aso caldera, southwest Japan (Fig. 1). Mount Aso is one of the largest active volcanoes on the Earth, with a caldera area of ~380 km^2. Activity initiated at the Aso volcanic cluster ~0.3 Myr ago with a large eruption that generated extensive pyroclastic flows. Four subsequent large, explosive eruptions resulted in the formation of Aso caldera (Ono and Watanabe 1985; Okubo and Shibuya 1993). Pyroclastic flows and volcanic ash from the caldera-forming eruption sequence covered a wide region of central Kyushu, including the study area. The basement rocks of the study area are mainly composed of Paleozoic metamorphic rocks, non- or weakly metamorphic eugeosynclinal rocks, Mesozoic granitic rocks, and marine sediments (Ono and Watanabe 1985).

In the southwest part of Aso caldera, the Hinagu–Futagawa fault zone, composed of the NNE–SSW- to NE–SW-striking Hinagu fault and NE–SW- to ENE–WSW-striking Futagawa fault, extends for ~81 km (Fig. 1b). Previous studies have shown that both the Hinagu and Futagawa faults are currently active, with recurrence intervals for large earthquakes of 3600–11,000 years for the Hinagu fault and 2600–8100 years for the Futagawa fault. The most recent event on the Hinagu fault occurred between 1200–1600 years BP, whereas the most recent large event on the Futagawa fault occurred 2200 years BP (Headquarters for Earthquake Research Promotion 2016). However, trench investigations occurred that were conducted after the 2016 Kumamoto earthquake reveal that: (1) prior to the 2016 earthquake, at least two morphogenic earthquakes occurred in the past ~2000 years on the Hinagu Fault, and four events in the past 4000–5000 years on the Futagawa Fault, suggesting an average late Holocene recurrence interval of 1000 years for morphogenic earthquakes within the HFFZ; and (2) the most recent event occurred in the period between AD ~850 and AD ~1150 AD; and (3) Tthe penultimate morphogenic event took place in the period between AD ~80 and AD ~200 AD (Lin et al. 2017). The 2016 Kumamoto earthquake occurred in the jog area between the Hinagu and Futagawa faults, ~30 km southwest of Aso caldera (Fig. 1b; Japan Meteorological Agency 2016; Geospatial Information Authority of Japan 2016). Historical and instrumental records show that >10 large earthquakes ($M \geq 6.0$) have occurred in the study area (Headquarters for Earthquake Research Promotion 2016). Seismic and geological data show that the Hinagu–Futagawa fault zone is currently active, and its earthquakes frequently cause severe damage, as exemplified by the 2016 Kumamoto main shock (Headquarters for Earthquake Research Promotion 2016).

References

Geospatial Information Authority of Japan (2016) Information concerning with the 2016 Kumamoto earthquake. http://www.jishin.go.jp/regional_seismicity/rs_katsudanso/f093_futagawa_hinagu/

Headquarters for Earthquake Research Promotion (2016) Evaluation of Heisei 28 (2016) Kumamoto earthquake. http://www.mext.go.jp/component/a_menu/science/detail/__icsFiles/afieldfile/2016/04/15/1285728_001_1.pdf

Lin A, Chen P, Satsukawa T, Sado K, Takahashi N, Hirata S (2017) Millennium recurrence interval of morphogenic earthquakes on the seismogenic fault zone that triggered the 2016 Mw 7.1 Kumamoto earthquake, SW Japan. Bull Seism Soc Am, in press. doi: 10.1785/0120170149

Japan Meteorological Agency (2016) Report on the Heisei 28 (2016) Kumamoto earthquake (No. 10). http://www.jma.go.jp/jma/menu/h28_kumamoto_jishin_menu.html

Okubo Y, Shibuya A (1993) Thermal and crustal structure of the Aso volcano and surrounding regions constrained by gravity and magnetic data. Japan J Volcanol Geotherm Res 55:337–350

Ono K, Watanabe K. (1985) Geological map of Aso Volcano, 1:50,000. Geological Map of Volcanoes 4. Geological Survey of Japan

A. Lin, *The 2016 Mw 7.1 Kumamoto Earthquake*, https://doi.org/10.1007/978-981-10-5855-4_2

3 Deformation Features of Coseismic Surface Rupture

3.1 Terminology

The term *coseismic surface rupture* is defined as a surface fracture produced by a current or large historic earthquake. The term is interchangeable with *surface earthquake fault* and *earthquake fault* in Japan, which is enhanced for the topographic morphology and geometry of surface fractures formed during large earthquakes [Research Group for Active Faults of Japan (RGAFJ) 1980, 1991]. In general, it is difficult to understand whether or not surface ruptures, including slope failures and landslides, are directly caused by seismogenic faulting or strong ground shaking during individual large earthquakes. In this book, to avoid any confusion regarding the terminology, we use the term *coseismic surface rupture* for the surface faults, fractures, cracks, and mole tracks that occurred during the 2016 Kumamoto earthquake, apart from when referring to distinct slope failures and landslides that occurred locally as defined by Lin (2017).

3.2 Outline of Coseismic Surface Rupture

Field investigations and analyses of aerial photographs and satellite images acquired immediately 1–3 days after the 2016 Kumamoto earthquake reveal that the 2016 M_w 7.1 Kumamoto earthquake produced a ~40-km-long surface rupture zone striking NE–SW on central Kyushu Island, Japan. Coseismic surface ruptures were characterized by shear faults, extensional cracks, and mole tracks, which mostly occurred along the preexisting NE–SW-striking Hinagu and Futagawa faults of the HFFZ in the southwest and central segments and newly identified faults in the northeast segment (Lin et al. 2016, 2017; Lin and Chiba 2017). Our results show that (i) the HFFZ triggered the 2016 Kumamoto earthquake and controlled the spatial distribution of coseismic surface ruptures and (ii) the southwest and central segments were dominated by right-lateral strike–slip movement with a maximum displacement of up to 2.5 m, accompanied by a minor vertical component. In contrast, the northeast segment was dominated by normal faulting with a maximum vertical offset of up to 1.75 m with a minor horizontal component that formed graben structures inside Aso caldera; (iii) coseismic rupturing initiated at the jog area between the Hinagu and Futagawa faults and then propagated northeastward into Aso caldera, where it terminated (Lin et al. 2016; Lin 2017).

3.3 Distribution Pattern of Coseismic Surface Rupture

Field investigations reveal that the 2016 M_w 7.1 Kumamoto earthquake produced a ~40-km-long surface rupture zone striking NE–SW in the central part of Kyushu Island, from the east side of Shimabara Bay in the southwest to Aso caldera in the northeast (Fig. 1). Based on structural features and distribution patterns of coseismic surface ruptures, the rupture zone can be divided into four segments, from southwest to northeast, which are the southwest, southwest–central (SW–central), northeast–central (NE–central), and northeast segments (Fig. 3).

3.3.1 Southwest and Southwest–Central Segments

The southwest segment is branched into two sub-rupture zones. One occurred mostly along the main fault trace of the Hinagu fault (called Zone-S1), along the topographic boundary between lowlands in the west and mountain slopes in the east, striking N10–30°E. The other (called Zone-S2) was distributed across the lowlands bounded by the Kasegawa River, striking N70–80°E, oblique to the trace of the Futagawa fault at an angle of 10–30° (Fig. 3). The ruptures of Zone-S1 were concentrated in a zone of width < 30 m (generally 3–10 m) along the trace of the Hinagu Fault. In contrast, the surface ruptures in Zone-S2 were dispersed across a zone >100 m wide. The SW–central segment occurred mostly along the main trace of the Futagawa fault,

A. Lin, *The 2016 Mw 7.1 Kumamoto Earthquake*, https://doi.org/10.1007/978-981-10-5855-4_3

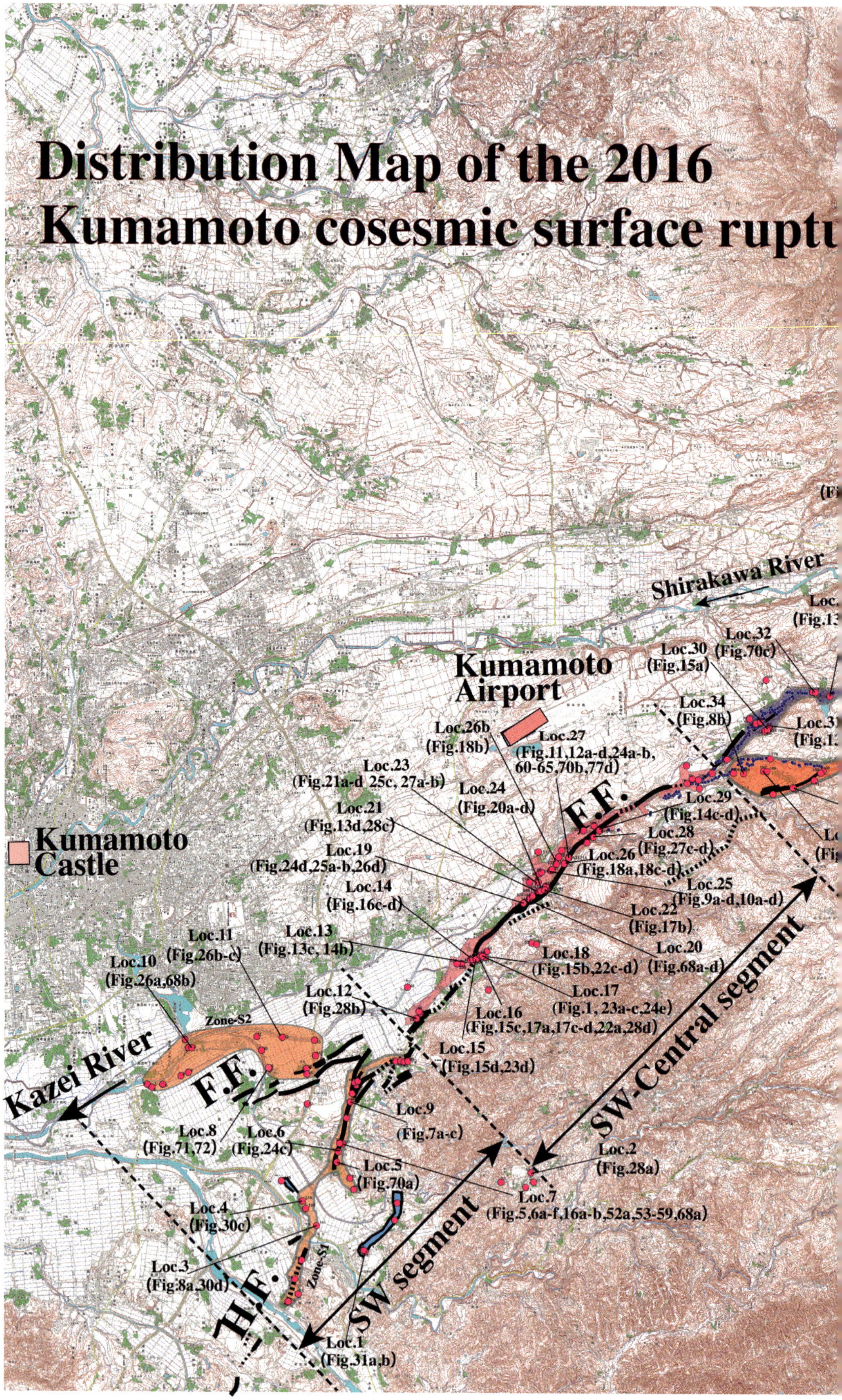

Fig. 3 Index map showing the distribution of 2016 coseismic surface ruptures (modified from Lin 2017). Active faults are cited from the Geospatial Information Authority of Japan (2016) and Headquarters for Earthquake Research Promotion (2016) Evaluation of Heisei 28 (2016) Kumamoto earthquake (http://www.mext.go.jp/component/a_menu/science/detail/__icsFiles/afieldfile/2016/04/15/1285728_001_1.pdf). Detail locations of figures with latitude and longitude are listed in Table 1

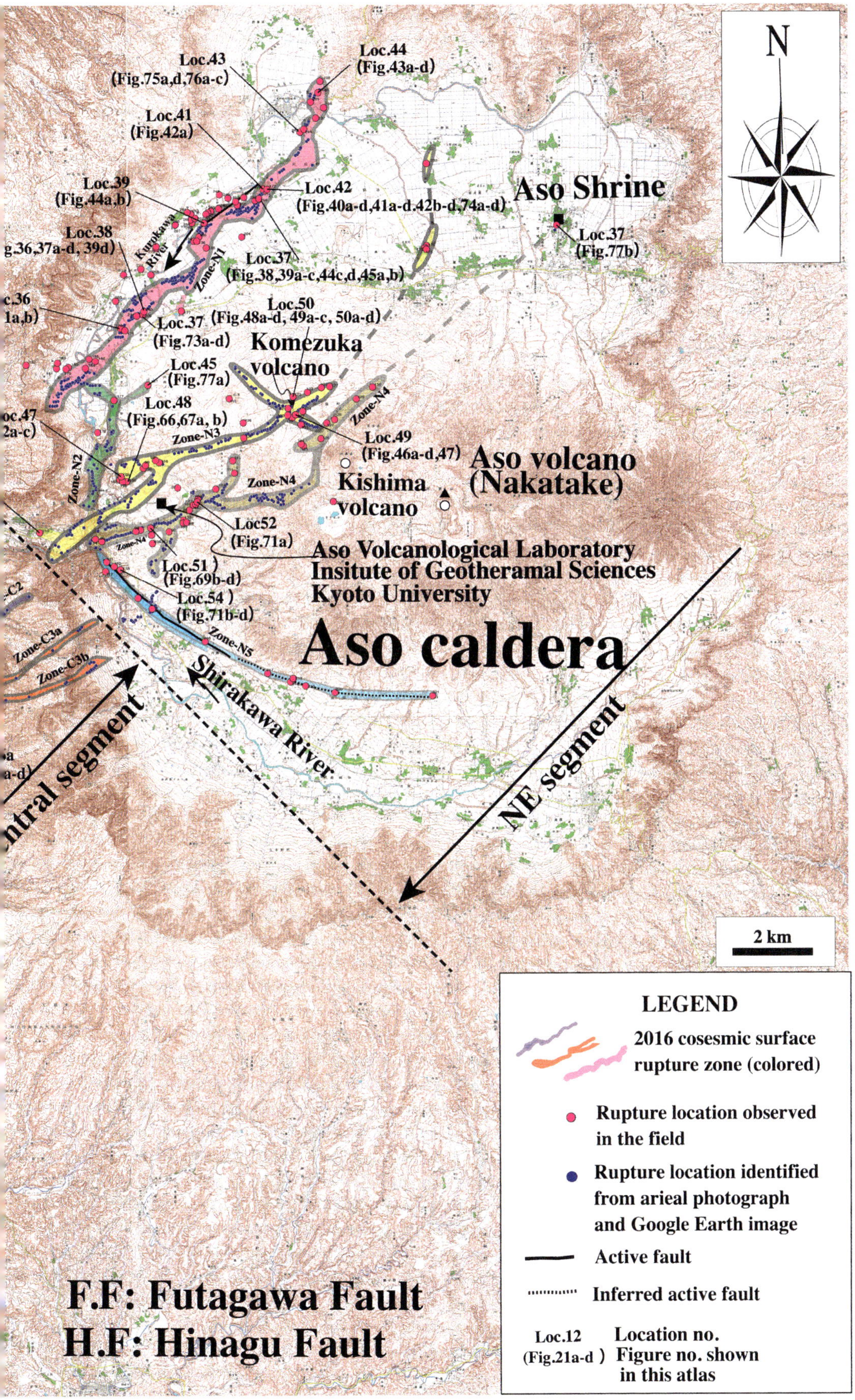

N
Loc.44
(Fig.43a-d)
Loc.43
(Fig.75a,d,76a-c)
Loc.41
(Fig.42a)
Loc.39
(Fig.44a,b)
Loc.42
(Fig.40a-d,41a-d,42b-d,74a-d)
Aso Shrine
Loc.37
(Fig.77b)
Loc.38
g.36,37a-d, 39d)
Kurokawa River
Zone-N1
Loc.37
(Fig.38,39a-c,44c,d,45a,b)
c.36
1a,b)
Loc.50
(Fig.48a-d, 49a-c, 50a-d)
Loc.37
(Fig.73a-d)
Komezuka volcano
Loc.45
(Fig.77a)
Loc.48
(Fig.66,67a, b)
oc.47
2a-c)
Zone-N4
Zone-N3
Loc.49
(Fig.46a-d,47)
Aso volcano
(Nakatake)
Zone-N2
Zone-N4
Kishima volcano
Loc52
(Fig.71a)
Aso Volcanological Laboratory
Insitute of Geotheramal Sciences
Kyoto University
Zone-N4
Loc.51
(Fig.69b-d)
Loc.54
(Fig.71b-d)
Zone-N5
Aso caldera
Zone-C3a
Zone-C3b
Shirakawa River
ntral segment
NE segment
2 km
LEGEND
2016 cosesmic surface rupture zone (colored)
Rupture location observed in the field
Rupture location identified from arieal photograph and Google Earth image
Active fault
Inferred active fault
Loc.12 (Fig.21a-d) Location no. Figure no. shown in this atlas
F.F: Futagawa Fault
H.F: Hinagu Fault

Table 1 Numbers of figures and detail latitude and longitude of locations shown in Fig. 3

Loc. no	Latitude(N)	Longitude(E)	Segment	Fig. No (shown in Fig. 3)
1	32.714203	130.803245	SW1	31a–b
2	32.732408	130.848805	SW1	28a
3	32.721185	130.789742	SW1	8a, 30d
4	32.726283	130.785886	SW1	30c, 68d
5	32.735121	130.795105	SW1	71a
6	32.737483	130.795599	SW1	24c
7	32.73895	130.796051	SW2	5, 6a–f, 16a–b, 52a, 53–59, 69a
8	32.75623	130.77697	SW2	72,73
9	32.748838	130.798854	SW2	7a–c
10	32.761221	130.754643	SW2	26a, 68c, 69b
11	32.763414	130.780333	SW2	26b–c
12	32.76794	130.81744	SW-Central	28b
13	32.780704	130.832511	SW-Central	13c, 14b
14	32.780892	130.833069	SW-Central	16c–d
15	32.781095	130.83382	SW-Central	15d, 23d
16	32.781257	130.834306	SW-Central	15c, 17a,17c, 17d, 22a, 28d
17	32.782144	130.835171	SW-Central	23a–c, 24e
18	32.78174	130.836192	SW-Central	15b, 22c, 22d
19	32.794156	130.847485	SW-Central	24d, 25a, 25b
20	32.795035	130.8485	SW-Central	26d
21	32.796402	130.850974	SW-Central	13d, 28c
22	32.79722	130.852975	SW-Central	17b
23	32.798598	130.850792	SW-Central	21a–d, 25c, 27a, 27b
24	32.796917	130.852465	SW-Central	20a–d
25	32.801787	130.853962	SW-Central	9a–d, 10a–d
26	32.481108	130.512812	SW-Central	18a, 18c, 18d
27	32.80471	130.85615	SW-Central	18b
28	32.804285	130.859022	SW-Central	11, 12a–d, 24a, 24b, 60–65, 71b, 78d
29	32.80919	130.865591	SW-Central	27c, 27d
30	32.809976	130.866699	SW-Central	14c, 14d
31	32.834052	130.914639	NE-Central	15a
32	32.834404	130.91494	NE-Central	13b
33	32.842551	130.926384	NE-Central	71c
34	32.841403	130.931819	NE-Central	13a, 29a–c
35	32.824825	130.913636	NE-Central	70a
36	32.823267	130.907237	NE-Central	8b
37	32.820644	130.921475	NE-Central	14a
38	32.83335	130.94244	NE1	32–34, 35a–d
39	32.943427	131.08015	NE1	51a, 51b
40	32.928333	131.00216	NE1	74a–d
41	32.929089	131.002137	NE1	36, 37a–d, 39d
42	32.948768	131.018493	NE1	44a, 44b
43	32.956078	131.036409	NE1	38,39a–c, 44c, 44d, 45a, 45b
44	32.942902	131.080159	NE1	42a
45	32.956435	131.036795	NE1	40a–d, 41a–d, 42b–d, 75a–d
46	32.969621	131.045665	NE1	76a, 76d, 77a–c
47	32.978214	131.050789	NE1	43a–d
48	32.911825	131.002783	NE2	78a
49	32.878738	130.973736	NE3	71d, 78c
50	32.891428	130.995914	NE3	73a–c
51	32.89031	130.997381	NE3	66, 67, 68a, 68b

(continued)

Table 1 (continued)

Loc. no	Latitude(N)	Longitude(E)	Segment	Fig. No (shown in Fig. 3)
52	32.905308	131.044064	NE3	46a–d, 47
53	32.909081	131.043627	NE3	48a–d, 49a–c, 50a–d
54	32.87935	131.003495	NE4	70b–d
55	32.885534	131.016587	NE4	72a
56	32.948179	131.116676	NE4	78b
57	32.870307	130.995683	NE5	72b–d

SW1 Southwest segment Zone-S1, *SW2* Southwest segment Zone-S2, *NE-Central* Northeast segment, *SW-Central* Southwest segment, *NE1* Northeast segment Zone-N1, *NE2* Northeast segment Zone-N2, *NE3* Northeast segment Zone-N3, *NE4* Northeast segment Zone-N4, *NE5* Northeast segment Zone-N5

striking N50–60°E, which developed along the topographic boundary between the lowlands formed by the westward flowing rivers and the southwestern slope of Mount Aso (Fig. 3). The ruptures were generally concentrated in a zone ranging from 2–3 m to ~100 m in width (typically 5–10 m). Locally, WSW–ENE- to E–W-striking surface ruptures with distinct shear faults occurred over a wide area that is 50–100 m from the NE–SW-striking rupture zone, which forms a conjugate rupture structure to the NE–SW-striking ruptures.

3.3.2 Northeast–Central Segment

The NE–central segment, striking NE–SW, mainly occurred on the southwestern slope of Mount Aso and comprises four subparallel rupture zones (called Zone-C1 to Zone-C3) (Fig. 3). Zone-C1 is distributed in the northern bank of the Shirakawa River, where numerous houses were mostly collapsed. The sub-rupture zone occurred on the lowlands and comprises mainly of extensional cracks. Zone-C2 is at the northeast extension of the SW-central segment, along the topographic boundary between the Shirakawa River valley and the slope of Mount Aso, where the Futagawa fault developed. Zone-C3 is on the southwestern slope of Mount Aso, 2–3 km east of Zone-C2, which is branched into two subparallel zones in the northeast part (Zone-C3a and Zone-C3b) (Fig. 3). It was difficult to access the rupture locations of Zone-C1 and Zone-C3 in the first 3 months after the earthquake, primarily due to earthquake damage to mountain roads. Therefore, most surface ruptures in this zone were identified from the high-resolution Google Earth images acquired on 16~18 April 2016 after the earthquake.

3.3.3 Northeast Segment

In contrast to the central and southwest segments, the northeast segment shows a relatively complicated distribution of surface rupture patterns in a wide area around the western and southern sides of Aso caldera. Based on the distribution and deformation features, this segment can be subdivided into five branch rupture zones (called Zone-N1 to Zone-N5), each with a different orientation (Fig. 3). Zone-N1, striking N40–60°E, is mainly composed of normal faults and extensional cracks that form graben structures (see below for details), extending ~10 km along the northwestern edge of Aso caldera (Fig. 3). Zone-N2 occurs along the southwestern edge of Aso caldera, striking N–S, and oblique to Zone-N1. Zone-N3 is the northeastern extension of Zone-C2 of the NE–central segment, which crosscuts the southwestern rim of Aso caldera and Komezuka cone (inside the caldera) with a conjugate geometric pattern of ruptures striking N50–60°E and N50–70°W, respectively (Fig. 3). Surface ruptures are also found in the area around the crater and foot of Komezuka cone in a doughnut-shaped pattern (see below for details). The ruptures of Zone-N1 and Zone-N3 terminated at the northeastern side, near the northern edge of the caldera. Zone-N4 shows more irregular geometric pattern than Zone-N1, which is locally bended and branched. This zone is subparallel to the general trend of Zone-N3, 0.5–3 km east of Zone-N3; crosscuts the southwest–northwestern side of Aso caldera, bounded by Kishima and Nakadake cones in the east and Komezuka cone in the west; and terminates at the northeastern edge of the caldera. Rupture in this zone was inferred to terminate near Aso Shrine in the northeast, which was completely destroyed by the earthquake. Zone-N5 lies on the southern slope of Nakadake cone along the southern edge of the caldera (Fig. 3). Coseismic surface ruptures were observed along a linear scarp striking N70–80°W and dipping south, which developed on alluvial fans formed from southward-flowing drainages in the northern side of the northwestward flowing Shirakawa River.

3.4 Slip Distribution of Coseismic Fault

Field measurements of coseismic displacements are plotted in Fig. 4, representing samples taken at 148 sites (Lin 2017). The maximum right-lateral displacements in the SW– and NE–central segments, respective are up to 2.45–2.50 m (Fig. 4). The deformation features of surface ruptures and slickenside striations observed in a wheat field (where the

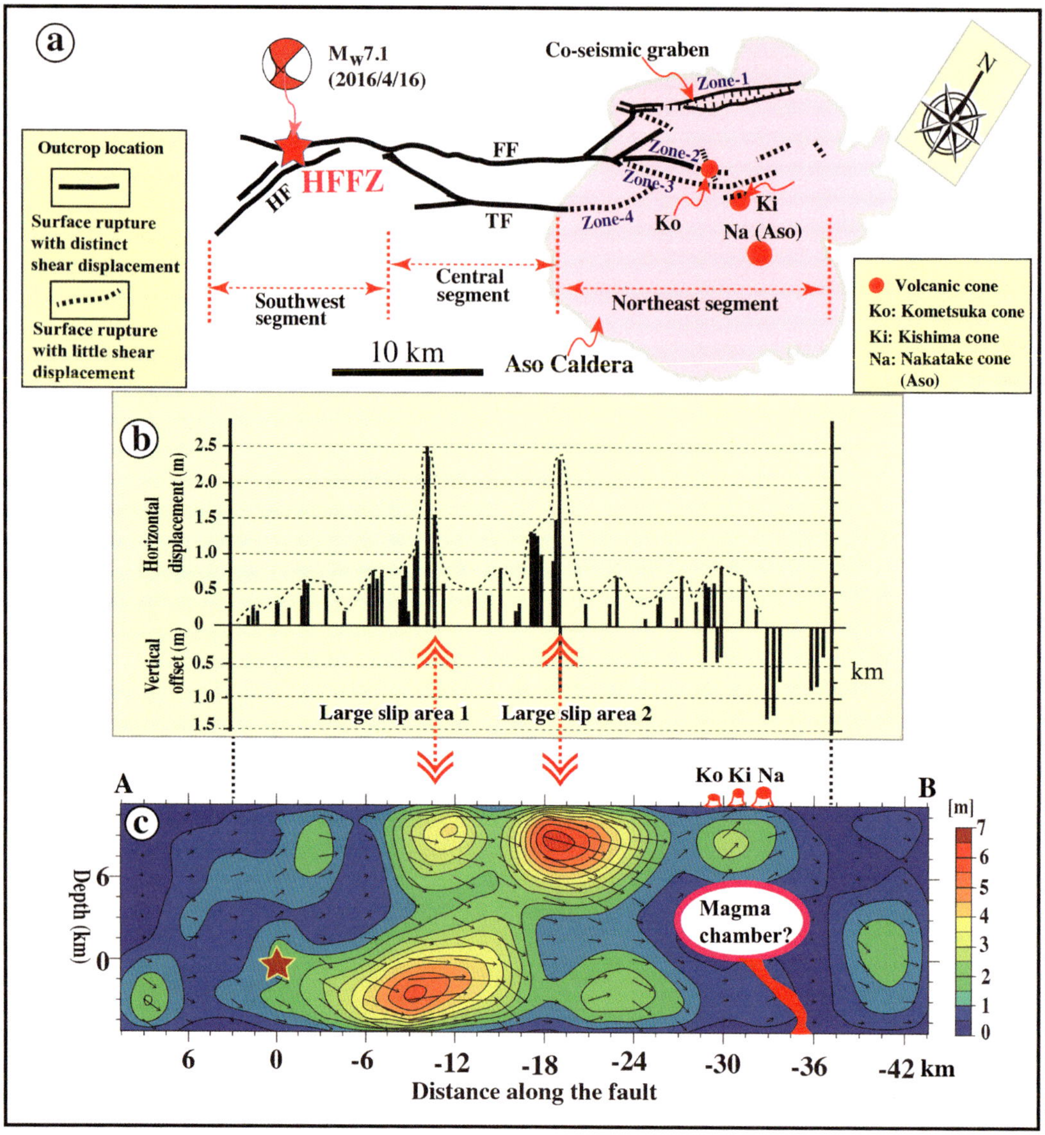

Fig. 4 Displacements and distribution and rupture traces of coseismic surface rupture zone (modified from Lin et al. 2016). (**a**) Distribution of coseismic displacements measured in-site. (**b**) Map showing the distribution of the coseismic surface rupture zone. (**c**) Seismic inversion results for seismic slip (in meters) on the seismogenic fault plane (modified from Koketsu et al. 2016) (see Fig. 1b for the surface location of profile A–B)

maximum displacement was observed) indicate a pure right-lateral strike–slip movement with little vertical component along the left-stepping echelon shear faults (Lin et al. 2016). On the mountain slope in the southwest of Aso caldera, right-lateral strike–slip displacements on mountain slope are often accompanied by distinct normal fault offset component along the coseismic shear faults. In both the SW– and NE–central segments, left-lateral displacements were observed locally along NW-striking shear faults (Fig. 4), in which the maximum offset amount is 0.9 m. In contrast, the northern segment inside Aso caldera is dominated by vertical displacement of up to 1.75 m with a minor horizontal component in both the left-lateral and right-lateral strike–slip faults that form conjugate fault structures (Fig. 4).

References

Geospatial Information Authority of Japan (2016) Information concerning with the 2016 Kumamoto earthquake. http://www.jishin.go.jp/regional_seismicity/rs_katsudanso/f093_futagawa_hinagu/

Headquarters for Earthquake Research Promotion (2016) Evaluation of Heisei 28 (2016) Kumamoto earthquake. http://www.mext.go.jp/component/a_menu/science/detail/__icsFiles/afieldfile/2016/04/15/1285728_001_1.pdf

Koketsu K, Kobayashi H, Miyake H (2016) Generation process of the 14th and 16th April 2016 Kumamoto earthquakes. http://taro.eri.u-tokyo.ac.jp/saigai/2016kumamoto/index.html#C

Lin A, Satsukawa T, Wang M, Mohammadi Asl Z, Fueta R, Nakajima F (2016) Coseismic rupturing stopped by Asovolcano during the 2016 Mw 7.1 Kumamoto earthquake, Japan. Science 354:869–875. doi:10.1126/science.aah4629

Lin A (2017) Structural features and seismotectonic implications of coseismic surface ruptures produced by the 2016 Mw 7.1 Kumamoto earthquake. J Seismol. 21:1079–1100. doi:10.1007/s10950-017-9653-5

Lin A, Chiba T (2017) Coseismic conjugate faulting structures associated with the 2016 Mw 7.1 Kumamoto earthquake, Japan. J Struct Geol 9:20–30. Open access article http://creativecommons.org/licenses/by-nc-nd/4.0/

Lin A, Chen P, Satsukawa T, Sado K, Takahashi N, Hirata S (2017) Millennium recurrence interval of morphogenic earthquakes on the seismogenic fault zone that triggered the 2016 Mw 7.1 Kumamoto earthquake, SW Japan. Bull Seismol Soc Am, in press. doi:https://doi.org/10.1785/0120170149

Research Group for Active Faults of Japan (RGAFJ) (1980) Active faults in Japan—sheet maps and inventories (revised edition), Univ. Tokyo Press, Tokyo, 363 pp (in Japanese with English summary)

Research Group for Active Faults of Japan (RGAFJ) (1991) Active faults in Japan—sheet maps and inventories (revised edition), Univ. Tokyo Press, Tokyo, 437 pp (in Japanese with English summary)

4 Structural Features of Coseismic Surface Ruptures

4.1 Southwest and Central Segments

Coseismic surface ruptures created different structural features in each of the four principal rupture segments (Lin 2017). The southwest segment is branched into two zones: Zone-S1 and Zone-2. Zone-S1 is mainly composed of distinct shear faults, left-stepping echelon cracks, and mole tracks occurred mostly along the main segment of the Hinagu fault (Fig. 3). Distinct shear faults striking N10–20°E and dipping 75–90° SE, subparallel to the general trend of the rupture zone, are dominated by right-lateral strike–slip movement. Horizontal slickenside striations observed on shear fault planes, marked by parallel lineations with some grooves and steps in unconsolidated clay, also show strike–slip-dominated movement. In contrast, Zone-S2 is mainly composed of extensional cracks and flexural structures along the southwestern segment of the Futagawa. The surface cracks are distributed over a wide area, and no distinct offset is observed. Liquefaction of sandy material occurred along the extensional cracks, in lowland areas near river channels, and was characterized by boiled sandy material along extensional cracks. Flexural structures formed in a field of vegetables as a waveform pattern, on which the extensional cracks duplicated (Lin et al. 2016; Lin 2017).

The SW–central segment occurred mostly along the main trace of the Futagawa fault, striking N50–60°E, which developed along the topographic boundary between the lowlands around the main drainages and the southwestern slope of Mount Aso (Fig. 3). The ruptures were generally concentrated in a zone ranging from 2 to 3 m to ~ 100 m in width (typically 5–10 m). Locally, WSW–ENE- to E–W-striking surface ruptures with distinct shear faults occurred over a wide area that is 50–100 m from the NE–SW-striking rupture zone, which forms a conjugate rupture structure to the NE–SW-striking ruptures (Lin and Chiba 2017).

The NE–central segment, striking NE–SW, mainly occurred on the southwestern slope of Mount Aso and comprises four subparallel rupture zones (called Zone-C1 to Zone-C3) (Fig. 3). Zone-C1 is distributed in the northern bank of the Shirakawa River, where numerous houses were mostly collapsed. The sub-rupture zone occurred on the lowlands and comprises mainly of extensional cracks. Zone-C2 is at the northeast extension of the SW–central segment, along the topographic boundary between the Shirakawa River valley and the slope of Mount Aso, where the Futagawa fault developed. Zone-C3 is on the southwestern slope of Mount Aso, 2–3 km east of Zone-C2, which is branched into two subparallel zones in the northeast part (Zone-C3a and Zone-C3b). It was difficult to access the rupture locations of Zone-C1 and Zone-C3, primarily due to earthquake damage to mountain roads. Therefore, most surface ruptures in this zone were identified from the high-resolution Google Earth images acquired on 18 April 2016 after the earthquake.

4.1.1 Coseismic Strike–Slip Fault (Figs. 5, 6, 7, 8, 9, 10, 11, 12, 13, 14, 15, 16, 17 and 18)

A. Lin, *The 2016 Mw 7.1 Kumamoto Earthquake*, https://doi.org/10.1007/978-981-10-5855-4_4

Fig. 5 Drone image showing the left-stepping tension cracks along the coseismic surface rupture zone in rice fields

Fig. 6 3D color shaded relief images and photographs showing the deformation features of coseismic surface rupture zone in rice fields. (**a–c**) 3D images acquired by Trimble TX5 3D laser scanner. (**d–f**) Right-lateral offset of paths (offset ~0.6 m)

e

f

a

b

Fig. 7 Right-lateral offset of a road. (**b, c**) Close-up view of (**a**) (offset ~0.3 m)

c

Fig. 8 Right-lateral offset of vegetable rows. (**a**) Right-lateral offset of vegetable rows at the southwestern end area of the coseismic surface rupture zone along the Hinagu fault (offset ~0.25 m). (**b**) Right-lateral offset of vegetable rows in the central segment along the Futagawa fault (offset ~0.15 m)

Fig. 9 Right-lateral offset of ditch slope and damaged parking area in a house garden. (**a, b**) Dextral offset ~1.3 m with a vertical component of ~0.30 m. (**c**) Right-lateral offset of ditch slope (offset ~0.9 m). (**d**) Right-lateral offset of parking area (offset ~1.1 m)

c

d

Fig. 10 Right-lateral offsets of a road, ditch, and fence wall. (**a–c**) Dextral offset ~1.3 m of the ditch (**a**), road (**b**), and wall fence (**c–d**)

c

d

Fig. 11 Drone image showing the right-lateral offset of fields at Masiki town

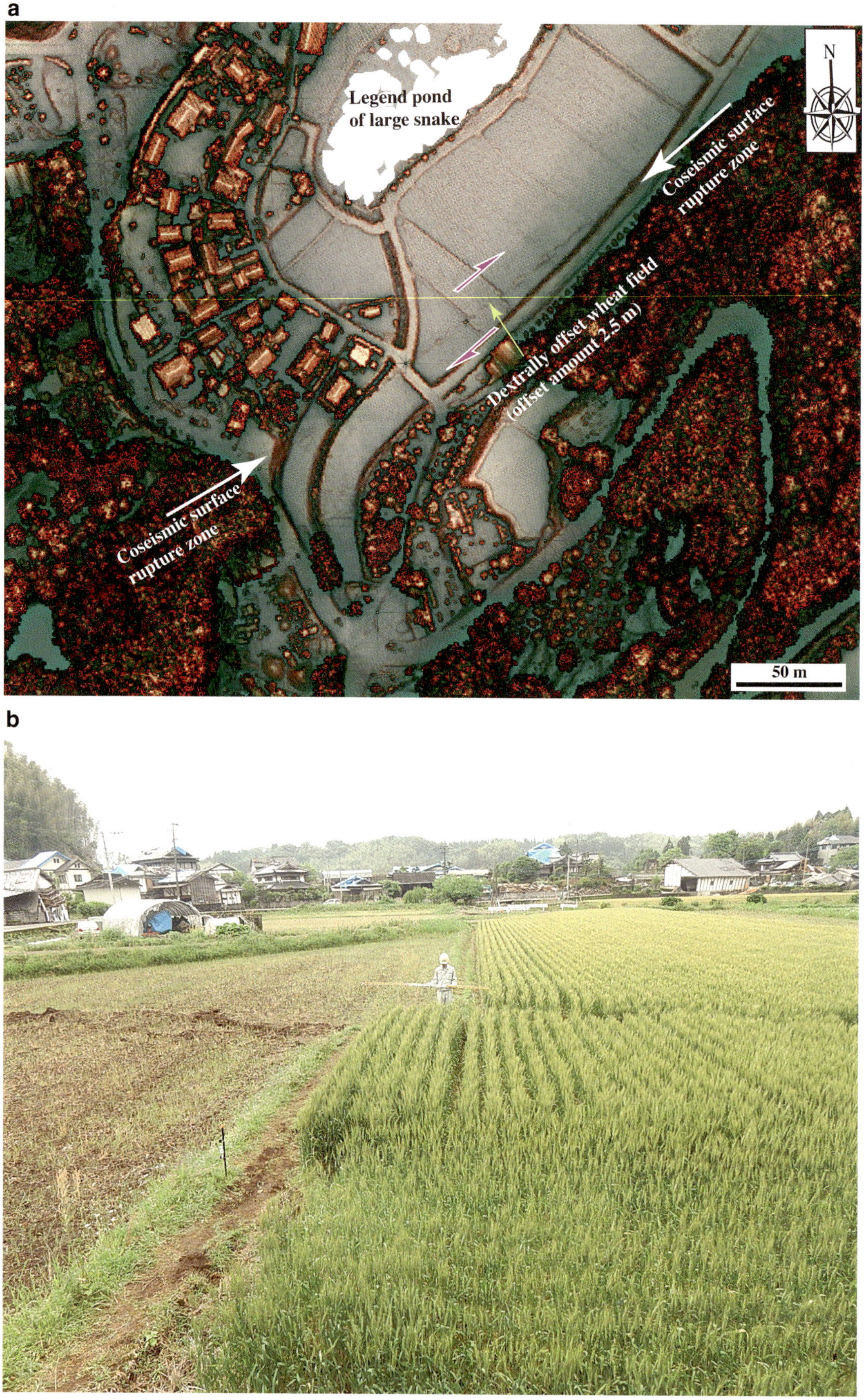

Fig. 12 Red relief image and photographs showing the right-lateral offset of fields. (**a**) Red relief image (image courtesy T. Chiba). (**b, c**) Right-lateral offset of wheat field. Photograph taken on 1 May 2016 (**d**) Photograph taken on 20 June 2016 after the wheat crops

c

d

a

b

Fig. 13 Right-lateral offset of roads. (**a**) Coseismic surface ruptures cut a road at Okirihata dam site. Dextral offset ~1.3 m (**b**), ~0.8 m (**c**) and 0.5 m (**d**)

c

d

a

b

Fig. 14 Right-lateral offset of roads and slope. (**a**) Offset ~0.5 m, (**b**) 0.8.m, (**c, d**) 0.7.m

c

d

a

b

Fig. 15 Right-lateral offset of fields, paths, and ditch. (**a**) Offset ~0.6 m, (**b**) 0.8 m, (**c, d**) 0.3 ~ 0.4 m

c

d

a

b

Fig. 16 Slickensides upon coseismic fault planes. (**a, b**) Coseismic fault plane developed within basement rocks (*sandstone*) and (**c, d**) unconsolidated surface soil layer. Striations developed on main fault planes indicate a strike–slip-dominated movement sense

c

d

a

b

Fig. 17 Right-lateral offset of fields. Drone images (**a–c**) and photograph (**d**) showing the right-lateral offset of fields. Coseismic surface ruptures preserved inside fields (**a–c**)

c

d

a

b

Fig. 18 Coseismic surface rupture occurred in weakly consolidated deposits. (**a**) Coseismic fault plane developed within volcanic deposits. (**b**) Coseismic surface rupture occurred in alluvial deposits. (**c**) Coseismic fault plane and slickensides developed upon a fault plane in alluvial deposits. (**d**) Close-up view of (**c**)

c

d

4.1.2 Coseismic Conjugate Riedel Shear Structure

Field investigations and analyses of LiDAR images reveal that the 2016 M_w7.1 Kumamoto earthquake produced a ~40-km-long surface rupture zone with a typical conjugate Riedel shearing fault structure along the preexisting right-lateral strike–slip (HFFZ). The conjugate Riedel shearing structure comprises two sets of coseismic shear fault zones that are oriented to NE–SW to ENE–WSW and WNW–ESE to E–W. The NE–SW- to ENE–WSW-trending shear fault zone is characterized by R Riedel shear structures with right-lateral strike–slip displacement of up to ~2.5 m, including left-stepping en echelon cracks (T-shear) and mole tracks (P-shear) (Fig. 32). In contrast, the WNW–ESE- to E–W--trending shear fault zone is dominated by R' Riedel shear structures with left-lateral displacement of up to1.3 m, including right-stepping en echelon tension cracks (T) and mole tracks (P), which are concentrated in a zone of <10 m within individual rupture zones (Lin and Chiba 2017). Our findings demonstrate that the coseismic conjugate Riedel shear faulting is mainly controlled by the preexisting active strike–slip faults of HFFZ under the present E–W compressive stress in the study area, associated with the ongoing penetration of the Philippine Sea Plate into the Eurasian Plate (Figs. 19, 20, 21, 22, 23, 24 and 25).

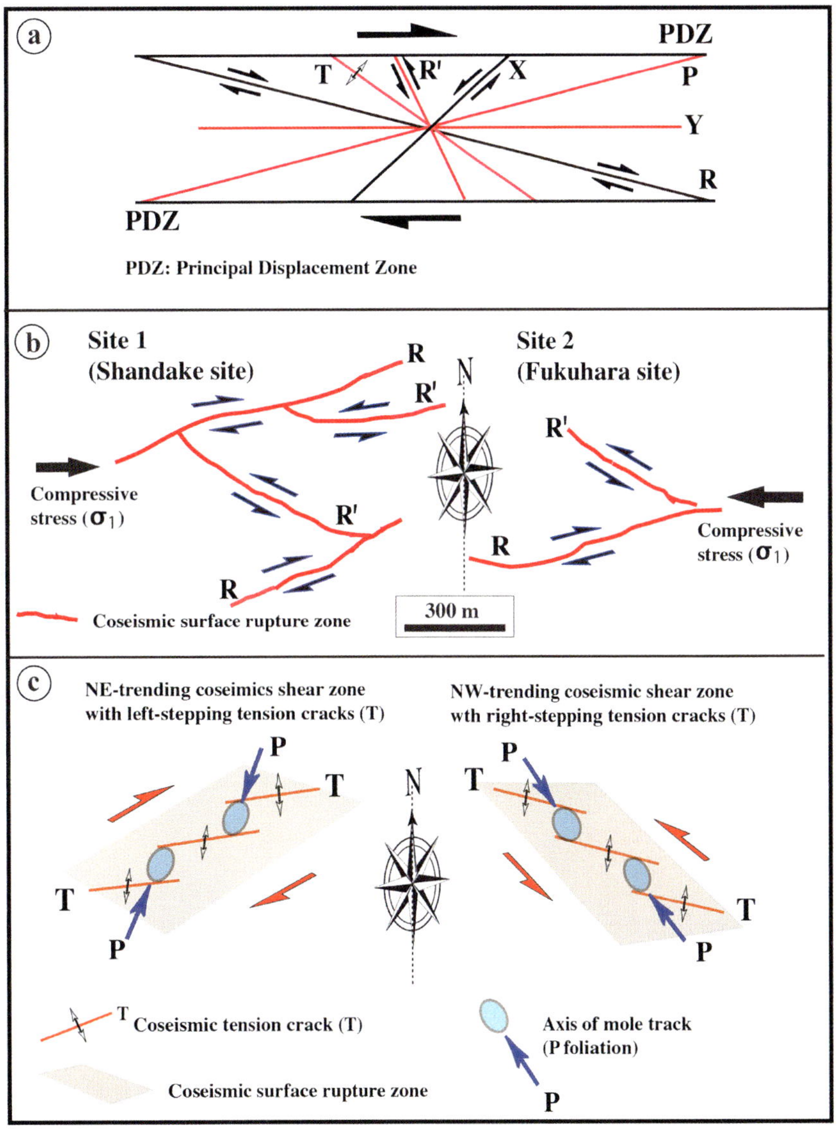

Fig. 19 Diagrams showing Riedel shear structural features (modified from Lin and Chiba [2017]). (**a**) The geometric features of an idealized Riedel shear zone. (**b**) The structural features of coseismic surface rupture zones mapped in typical areas. (**c**) Tension cracks (T) and mole tracks (P) are developed within the ENE-trending and WNW-trending coseismic Riedel shear zones

a

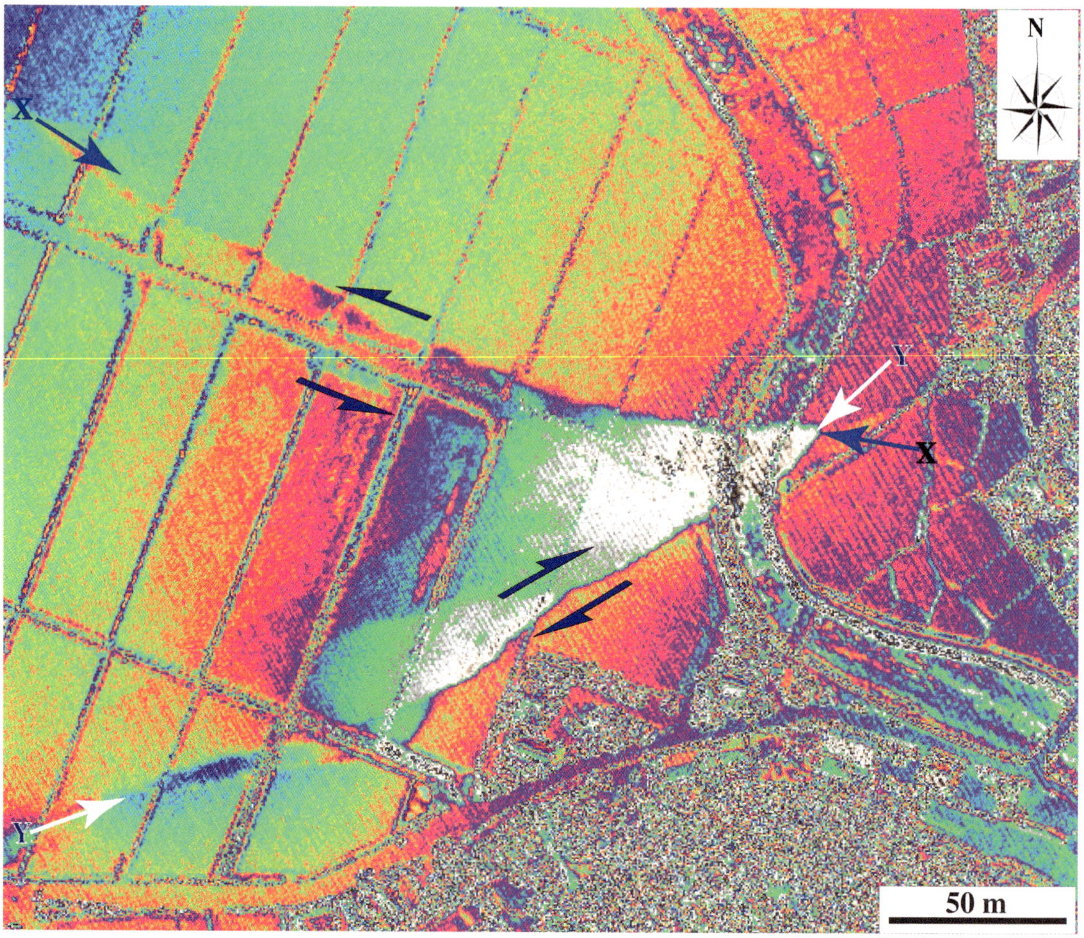

b

Fig. 20 LiDAR and drone images showing structural features of coseismic conjugate faults at Shantake village. (**a**) Right-lateral strike–slip displacements occurred along Y fault striking northeast–southwest, while left-lateral strike–slip displacements occurred along X fault striking northwest–southeast. (**b, c**) Drone images and (**d**) photograph showing right-lateral offset of river channel and the fields (offset ~0.9 m)

c

d

a

b

Fig. 21 Strike–slip offset of rice field rows. (**a–c**) Left-lateral offset of rice field rows along WNW-ESE striking rupture zone (offset amount ~ 0.3 m). (**d**) Right-lateral offset of rice field rows (offset ~0.6 m)

c

d

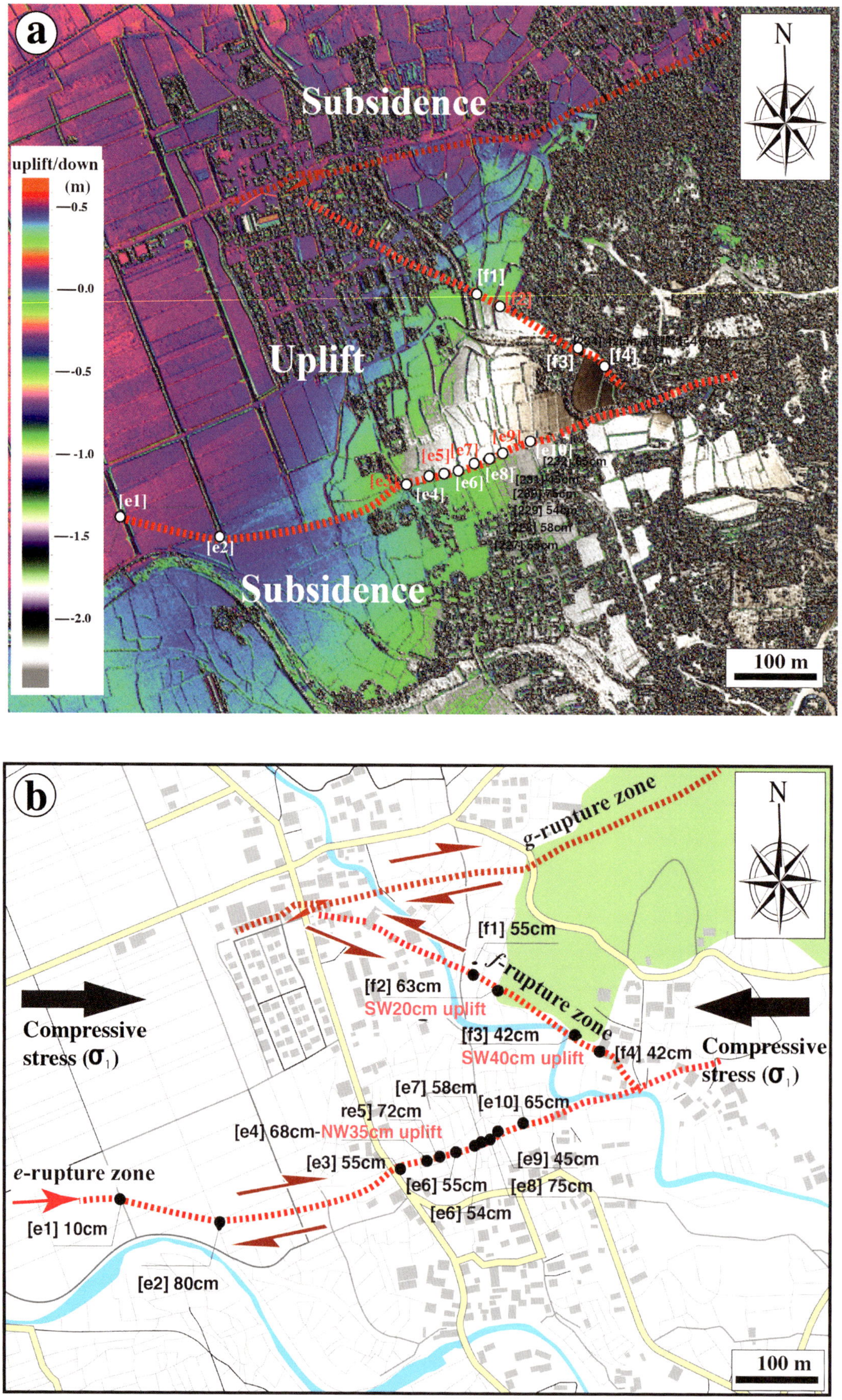

Fig. 22 LiDAR image and photographs showing structural features of coseismic conjugate faults at Fukuhara village (modified from Lin and Chiba [2017]). (**a, b**) Right-lateral strike–slip displacements occurred along R fault striking northeast–southwest, while left-lateral strike–slip displacements occurred along R' fault striking northwest–southeast. Conjugate ruptures occurred on a road (**c**) and vegetable field (**d**)

c

d

a

b

Fig. 23 Strike–slip offset of a road and field paths. (**a, b**) Left-lateral offset of a road and field. (**c**) Left-lateral offset of a field path (offset ~0.3 m). (**d**) Right-lateral offset of a field path (offset ~0.4 m)

c

d

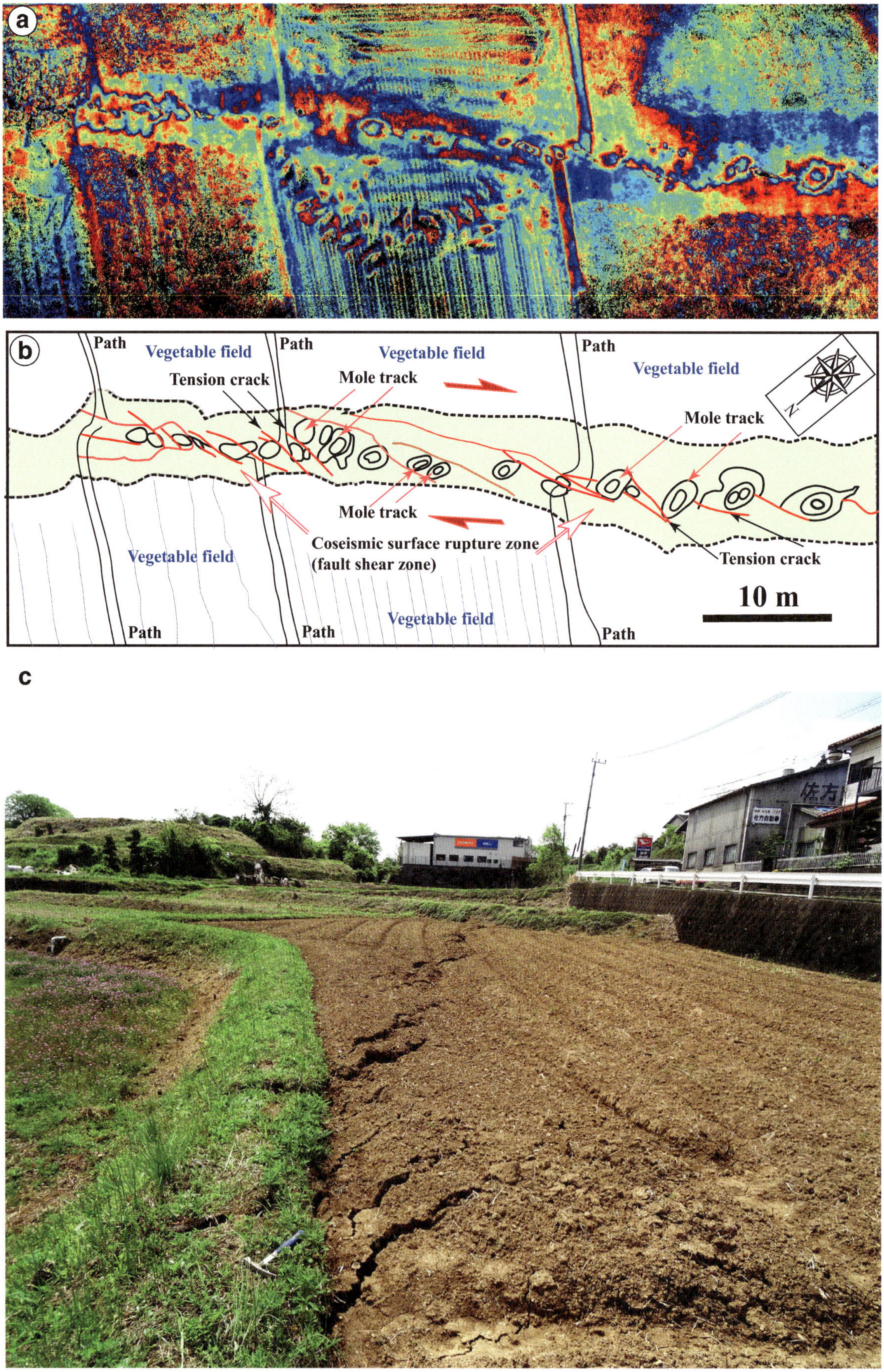

Fig. 24 3D color shaded relief image and photographs showing Riedel shear structures of the coseismic surface rupture zone (modified from Lin and Chiba [2017]). **(a)** Color shaded relief image acquired by Trimble TX5 3D laser scanner. **(b)** Sketch of (a). **(c–e)** Left-stepping en echelon tension cracks occurred along the NE–SW-striking coseismic surface rupture zone

d

e

a

b

Fig. 25 Coseismic en echelon tension cracks. (**a**) Left-stepping tension cracks occurred along the NE–SW-striking coseismic surface rupture zone. (**b, c**) Right-stepping tension cracks along the WNW–ESE-striking rupture zone

c

4.1.3 Coseismic Flexures and Mole Track

Axes of mole track structures are oriented in a right-stepping geometric pattern, indicating orientation of P foliations, while tension cracks show a left-stepping geometric pattern, indicating en echelon orientation of T-shears. The mole tracks are generally cut by the tension cracks (T-shear), indicating a formation timing of P- and T- shears. The long axes of mole track structures are generally oblique to the general trend of the coseismic surface rupture zone, and they form left-stepping en echelon patterns. The mole track structures (P-foliation) are generally found in the pinched and/or jog areas between the left-stepping tension cracks (T-shear), reflecting the compressive nature of the jog areas within the right-lateral strike–slip surface rupture zone. Morphologically, most of the mole track structures show an angular ridge pattern and some bulge patterns, similar to those produced during the 2001 M_w 7.8 Kunlun earthquake along the typical right-lateral strike–slip Kunlun Fault (Lin et al. 2004) (Figs. 26, 27 and 28).

a

b

Fig. 26 Flexural structures occurred along the coseismic surface rupture zone. (**a**) Flexure structure occurred at a vegetable field. (**b, c**) A road folded as a waveform. (**d**) Flexural structure occurred at a rice field

c

d

a

b

Fig. 27 Mole track structures occurred at roads. (**a, b**) Mole track along the WNW–ESE-striking coseismic surface rupture zone. (**c, d**) Mole track occurred on a road along the ENE–WSW-striking coseismic surface rupture zone

c

d

Fig. 28 Mole track structures occurred at a road and field path. (**a–c**) Mole track structure occurred at a road. (**d**) Mole track structure occurred at a field path along the ENE–WSW-striking coseismic surface rupture zone

c

d

4.1.4 Earthquake-Induced Liquefaction (Figs. 29, 30 and 31)

a
b

Fig. 29 Earthquake-induced liquefaction that occurred inside a dam. (**a**) Overview of the Okirihata dam that was damaged by the coseismic surface rupture zone. (**b, c**) Sand boils occurred inside the dam

c

Fig. 30 Earthquake-induced sand boils. (**a–c**) Sand boils occurred in association with earthquake-induced liquefaction that occurred within extensional cracks in vegetable fields. (**d**) Sand boils occurred in a lowland of river bank

c

d

Fig. 31 Ground subsidence caused by earthquake-induced liquefaction. (**a**) Manhole uplift along the extensional crack. (**b**) Close-up view of (**a**)

4.1.5 Coseismic Surface Ruptures Along Newly Identified Tawarayama Fault (Figs. 32, 33, 34 and 35)

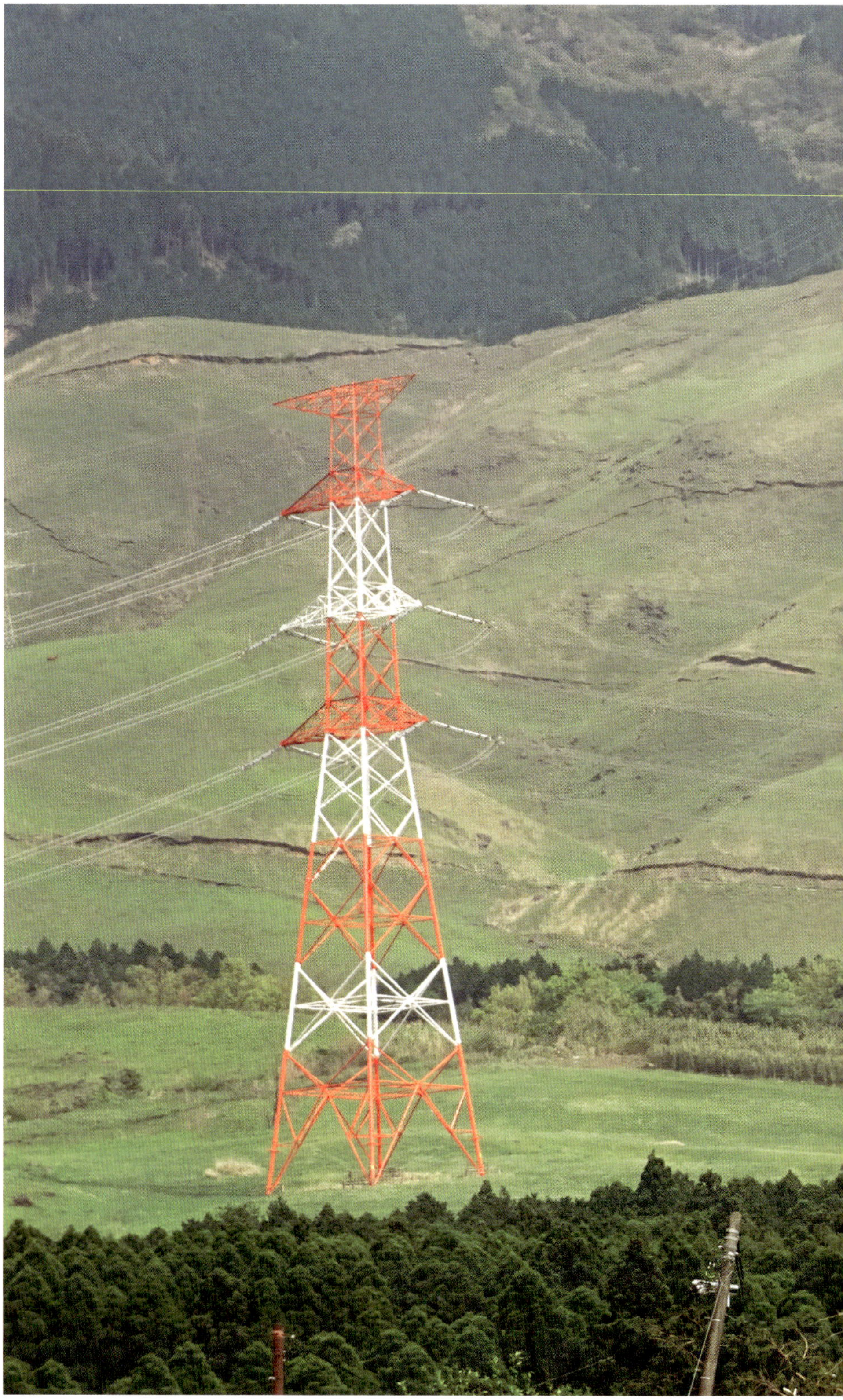

Fig. 32 All view of coseismic surface ruptures occurred on the slope of Tawarayama Mt

Fig. 33 Coseismic surface ruptures occurred on a slope of the Tawarayama Mt. (**b**) Close-up view of (**a**). (**c, d**) Close-up view of en echelon coseismic surface ruptures shown in (**a**) and (**b**)

c

d

a

b

Fig. 34 Coseismic shear faults. (**a–c**) Continuous coseismic shear faults occurred on the Tawarayama Mt. slope. (**d**) Coseismic shear fault plane exposed in pyroclastic flow deposits on a mountain slope

c

d

Striation
Striation
Striation
Striation
Striation
Striation

a

b

Fig. 35 Right-lateral strike–slip displacement of a gully and path on Tawarayama Mt. (**a**) Gully offset ~2.4 m. (**b**) Path offset ~1.7 m. (**c, d**) Coseismic surface ruptures without distinct vertical offset developed in pyroclastic flow deposits

c

d

4.2 Northeast Segment Inside Aso Caldera

4.2.1 Coseismic Graben Structure (Figs. 36, 37, 38, 39, 40, 41, 42, 43, 44 and 45)

Fig. 36 Drone image showing a coseismic graben structure developed inside Aso caldera

Fig. 37 Coseismic fault scarps of a graben structure and slickenside. (**a, b**) Coseismic fault scarp developed along one side of a graben structure (vertical offset up to ~1.7 m). (**c**) Slickenside on a fault plan of (**b**). (**d**) Drone image showing a graben structure where a pond formed

c

d

Fig. 38 Coseismic flexure structures formed in rice fields. Drone image showing a flexure feature of coseismic surface ruptures occurred in rice fields that are folded as waveform

a

b

Fig. 39 Drone images showing a coseismic graben structure formed in rice fields. (**a**) Rice fields folded as waveforms. (**b**) A graben structure formed in rice fields where a temporary pond formed. (**c**) Close-up view of (**b**). (**d**) Ground subsidence of up to ~0.4 m occurred in a rice field where a temporary pond formed

c

d

a

Fig. 40 A coseismic graben structure formed in a residential area. (**a**) Google Earth image showing a coseismic graben structure. (**b**) Drone image showing the coseismic surface ruptures formed a graben structure. (**c**) A graben structure formed on a road that was repaired immediately after the earthquake. Red roof houses located inside the graben, which are the same as that shown in (**a**) and (**b**)

b

c

Fig. 41 Coseismic normal fault scarps. (**a, b**) A normal fault scarp developed along one side of a graben structure (vertical offset ~1.3 m). (**c, d**) Vertically offset road (offset ~1.3 m)

c

d

a

b

Fig. 42 Coseismic graben structure and normal fault scarps. (**a**) Drone images showing a graben structure formed in rice fields. (**b**) A normal fault scarp developed along one side of a graben structure. (**c, d**) A normal fault scarp developed in a house garden (vertical offset ~1.2 m)

c

d

a

b

Fig. 43 Coseismic surface ruptures formed on roads and fields. (**a**) A coseismic graben structure formed in fields and roads. (**b**) A flexure structure formed in a rice field that was folded. (**c**) A temporary pond formed inside a coseismic graben structure. (**d**) Vertical offset of a road (offset ~0.4 m)

c

d

a

b

Fig. 44 Coseismic normal fault scarp along a graben structure. (**a**) A coseismic normal fault scarp cut a road and fields. (**b**) Close-up view of (a). (**c**) Drone image showing a coseismic graben structure formed in vegetable fields where a temporary pond formed. (**d**) Close-up of (**c**)

c

d

Fig. 45 Drone image showing en echelon coseismic surface ruptures developed in fields. (**b**) Close-up view of (**a**)

4.2.2 Coseismic Ruptures Crosscut Volcano Cones (Figs. 46, 47, 48, 49, 50 and 51)

a

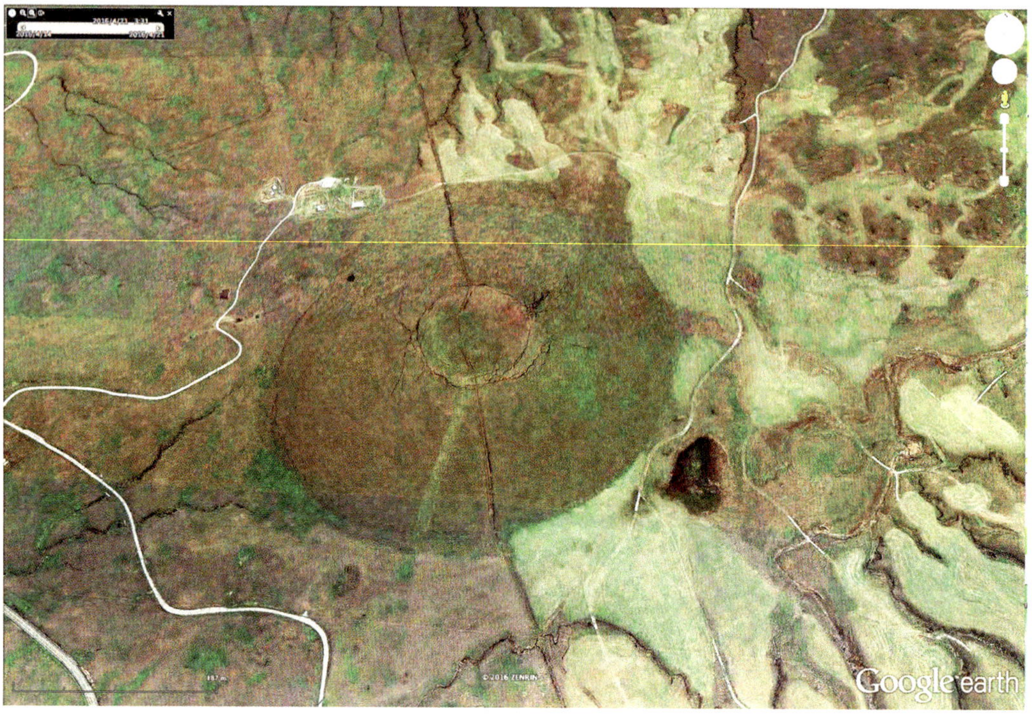

b

Fig. 46 Coseismic surface ruptures crosscut the Komezuka cone. (**a**) Google Earth image acquired on 16 April 2016 showing coseismic surface ruptures occurred in the area around the Komezuka volcano. (b) Drone image showing an overview of coseismic surface ruptures occurred on the *top* and *foot* of the Komezuka cone (acquired on 16 Nov. 2016). (**c**) Close-up view of (**b**). (**d**) Aerial photograph acquired on 18 April 2016 (copyright permission from Mainichi Newspapers)

c

d

Fig. 47 A doughnut-shaped coseismic surface ruptures occurred on the *top* of the Komezuka cone. Drone image acquired on 16 Nov. 2016

Fig. 48 Coseismic surface ruptures occurred on the Western slope of the Kishima cone. (**b**) Close-up view of (**a**). (**c**) Subparallel coseismic surface ruptures occurred on the slope. (**d**) Close-up view of (**c**)

c

d

Fig. 49 Coseismic surface ruptures occurred on the *top* of the Komezuka cone. (**a–c**) Coseismic tensional cracks occurred on the *top* of the Komezuka cone

c

a

b

Fig. 50 Coseismic surface ruptures occurred on the *top* and *foot* of the Komezuka cone. (**a, b**) Left-lateral offset of a linear soil bulge (offset 0.6 m). (**c, d**) Coseismic surface ruptures occurred around the foot of the Komezuka cone

c

d

a

b

Fig. 51 Coseismic surface ruptures occurred at the northeasternmost end site of the 2016 Kumamoto rupture zone in the northwestern edge of Aso caldera. (**a**) A continuous coseismic crack occurred in a rice field. (**b**) A temporary pond formed in a rice field along the coseismic surface rupture zone

References

Lin A (2017) Structural features and seismotectonic implications of coseismic surface ruptures produced by the 2016 Mw 7.1 Kumamoto earthquake. J Seismol 21:1079–1100. doi:10.1007/s10950-017-9653-5

Lin A, Chiba T (2017) Coseismic conjugate faulting structures associated with the 2016 mw 7.1 Kumamoto earthquake, Japan. J Struct Geol 9:20–30. Open access article http://creativecommons.org/licenses/by-nc-nd/4.0/

Lin A, Guo J, Fu B (2004) Co-seismic mole-track structures produced by the 2001 Ms 8.1 central Kunlun earthquake, China. J Struct Geol 26:1511–1519

Lin A, Satsukawa T, Wang M, Mohammadi Asl Z, Fueta R, Nakajima F (2016) Coseismic rupturing stopped by Aso volcano during the 2016 mw 7.1 Kumamoto earthquake, Japan. Science 354:869–875. doi:10.1126/science.aah4629

Trench Investigations on Seismogenic Faults

5

To understand the seismotectonic history of the HFFZ, including the recurrence interval of morphogenic earthquakes, slip rate, and source fault characteristics of the 2016 Kumamoto earthquake, we conducted paleoseismic studies during the past 1 year after the earthquake, including fieldwork on the 2016 coseismic surface ruptures and related active faults. Two trenches were excavated across the coseismic rupture zone at the Takagi and Kamijin sites corresponding to the maximum observed strike–slip displacements of the Hinagu fault and Futagawa fault (Lin et al. 2017).

Field investigations related to the Kumamoto earthquake, trench excavations across the Hinagu and Futagawa faults, and radiocarbon dating results reveal that (1) prior to the 2016 earthquake, at least two morphogenic earthquakes occurred in the past ca. 2000 years on the Hinagu fault and four events in the past 4000–5000 years on the Futagawa fault, suggesting an average late Holocene recurrence interval of 1000 years for morphogenic earthquakes within the HFFZ; (2) the most recent event occurred between AD 1000 and 1400; (3) the penultimate paleoseismic event took place at ca. 2000 year BP. The average strike–slip rates are estimated to be 0.5–0.7 mm/year for the Hinagu fault and 1.7–2.7 mm/year for the Futagawa fault (Lin et al. 2017). These results contradict previous studies that estimate recurrence intervals for morphologic earthquakes of 3600–11,000 years and 8000–26,000 years on the target segments of the Hinagu and Futagawa faults, respectively (Headquarters for Earthquake Research Promotion 2016). Main structural features of seismogenic faults revealed by two trenches are shown by images and sketches in this book (Figs. 52, 53, 54, 55, 56, 57, 58, 59, 60, 61, 62, 63, 64 and 65).

A. Lin, *The 2016 Mw 7.1 Kumamoto Earthquake*, https://doi.org/10.1007/978-981-10-5855-4_5

a

Fig. 52 3D color relief images of trenches. (**a**) Takagi trench. (**b**) Kamijin trench

b

Fig. 52 (continued)

Fig. 53 Drone image showing an overview of the Takagi trench site. A coseismic surface rupture zone occurred along the preexisting fault scarp of the Hinagu fault, where field paths were dextrally offset by 0.6 m. A trench was excavated immediately 3 months after the 2016 Kumamoto earthquake at this site

a

b

Fig. 54 Trenching scene at Takagi site. (**a, b**) Drone images showing a trenching scene. (**c, d**) Close-up view of trenching

c

d

Fig. 55 3D color relief image showing an overview of the trench walls exposed at the Takagi trench. Upside to the west

Fig. 56 Exposure walls of the Takagi trench. (**a**) North wall. (**b**) South wall

a

Fig. 57 Close-up view of faults exposed at Takagi trench. (**a**) North wall. (**b**) South wall

b

a

Fig. 58 Faults exposed at base wall of the Takagi trench. (**b**) Close-up view of (**a**)

b

Fig. 59 Exposure walls and sketches of the Takagi trench. (**a**) North wall. (**b**) South wall. (**c**) Close-up view of (**b**). (**d, e**) Sketches of (**a, b**)

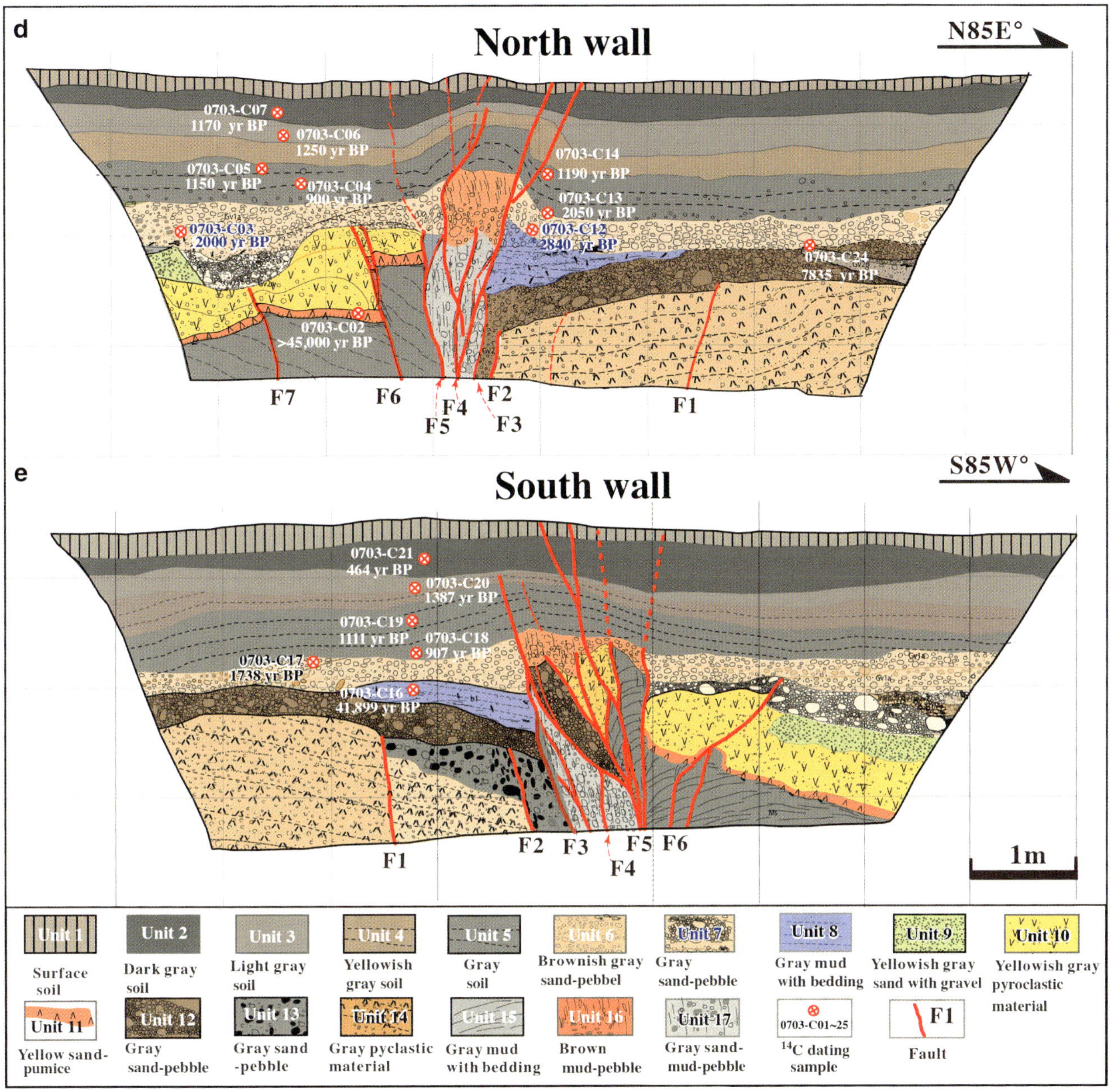

d
North wall
N85E°
0703-C07
1170 yr BP
0703-C06
1250 yr BP
0703-C05
1150 yr BP
0703-C04
900 yr BP
0703-C14
1190 yr BP
0703-C13
2050 yr BP
0703-C03
2000 yr BP
0703-C12
2840 yr BP
0703-C24
7835 yr BP
0703-C02
>45,000 yr BP
F7
F6
F4
F5
F2
F3
F1
e
South wall
S85W°
0703-C21
464 yr BP
0703-C20
1387 yr BP
0703-C19
1111 yr BP
0703-C18
907 yr BP
0703-C17
1738 yr BP
0703-C16
41,899 yr BP
F1
F2
F3
F4
F5
F6
1m
Unit 1 Surface soil
Unit 2 Dark gray soil
Unit 3 Light gray soil
Unit 4 Yellowish gray soil
Unit 5 Gray soil
Unit 6 Brownish gray sand-pebbel
Unit 7 Gray sand-pebble
Unit 8 Gray mud with bedding
Unit 9 Yellowish gray sand with gravel
Unit 10 Yellowish gray pyroclastic material
Unit 11 Yellow sand-pumice
Unit 12 Gray sand-pebble
Unit 13 Gray sand -pebble
Unit 14 Gray pyclastic material
Unit 15 Gray mud with bedding
Unit 16 Brown mud-pebble
Unit 17 Gray sand-mud-pebble
0703-C01~25 14C dating sample
F1 Fault

Fig. 60 Drone image showing an overview of the Kamijin trench site. Field paths were dextrally offset by ~2.5 m

a

b

Fig. 61 Trenching at the Kamijin site. (**a**) Drone image showing an overview of trenching. (**b, c**) An overview of the trenching site. (**d**) An overview of an exposure wall during heavy raining

c

d

Fig. 62 3D color relief image showing an overview of the trench walls exposed at the Kamijin trench

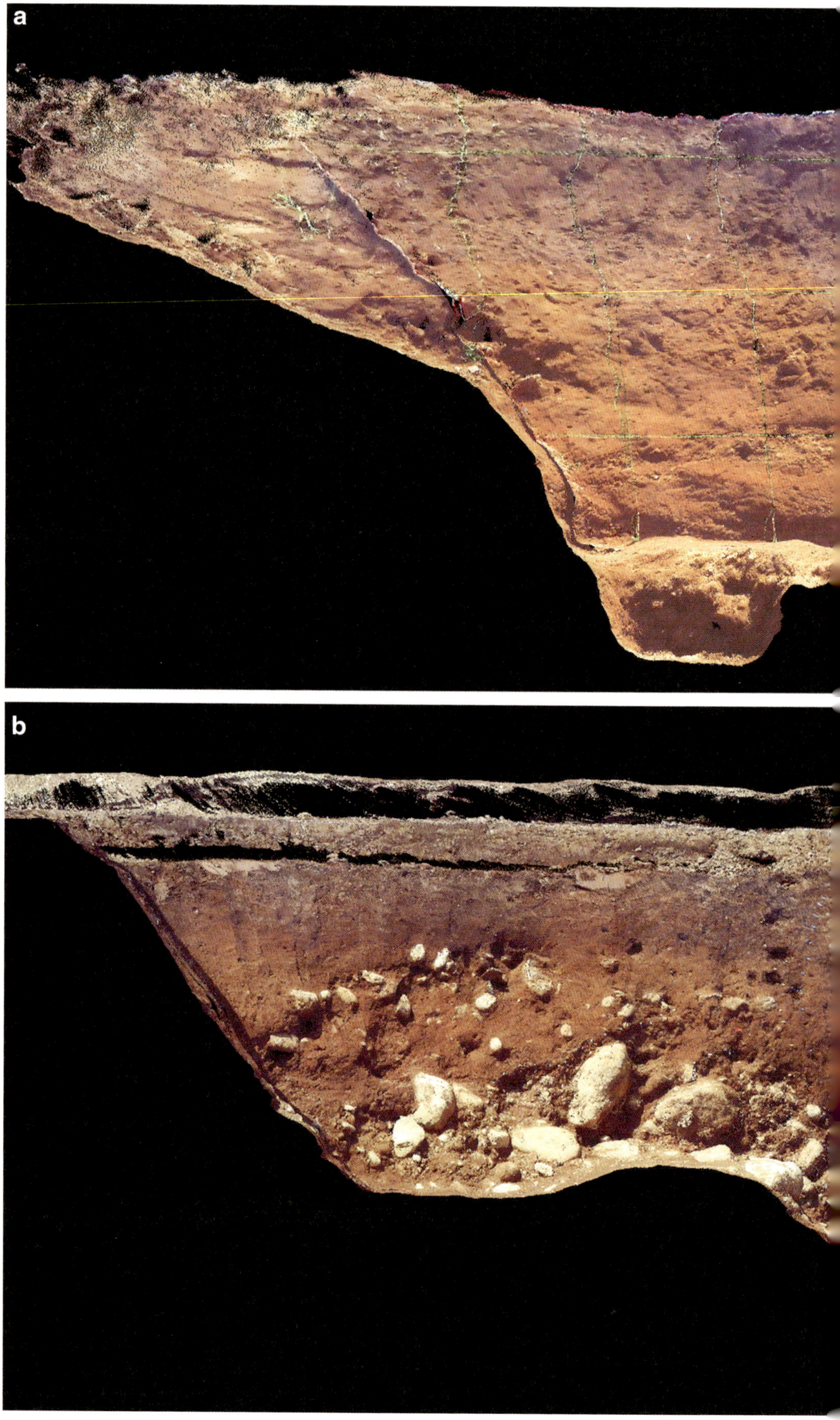

Fig. 63 Exposure walls of the Kamijin trench. (**a**) Northeast wall. (**b**) Southwest wall

a

Fig. 64 Injection vein of surface soil and striations. (**a**) Gray surface soil injected in alluvial deposits. (**b, c**) Striations developed upon a main fault plane exposed in the southwest wall of Kamijin trench

b

c

Fig. 65 Exposure walls of the Kamijin trench. (**a**) Northeast wall. (**b**) Southwest wall. (**c**) Base wall. (**d, e**). Sketches of exposure walls shown in (**a**) and (**b**)

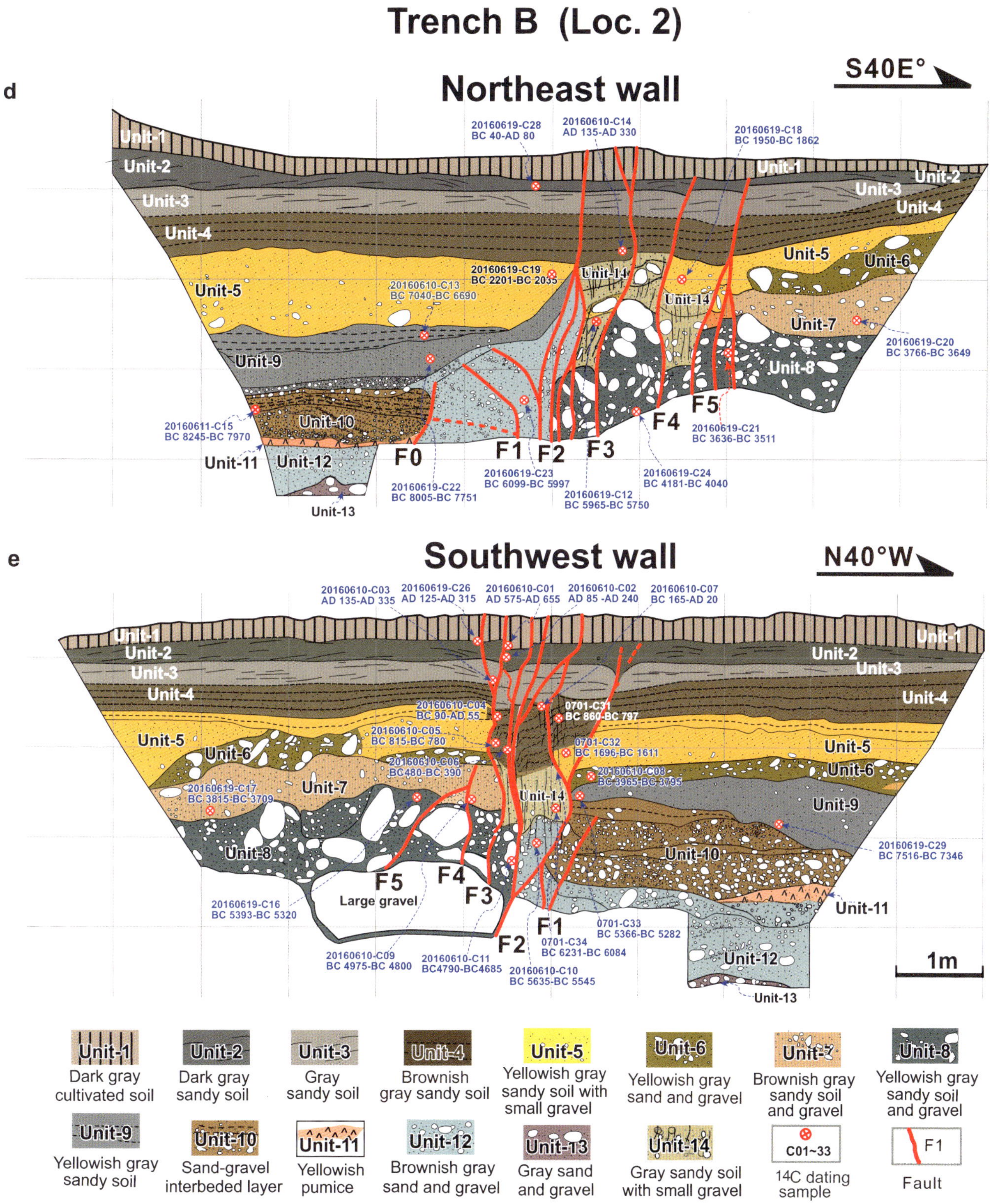
Trench B (Loc. 2)
d
Northeast wall
S40E°
Unit-1
Unit-2
Unit-3
Unit-4
Unit-5
Unit-6
Unit-7
Unit-8
Unit-9
Unit-10
Unit-11
Unit-12
Unit-13
Unit-14
20160619-C28 BC 40-AD 80
20160610-C14 AD 135-AD 330
20160619-C18 BC 1950-BC 1862
20160619-C19 BC 2201-BC 2035
20160610-C13 BC 7040-BC 6690
20160619-C20 BC 3766-BC 3649
20160611-C15 BC 8245-BC 7970
20160619-C21 BC 3636-BC 3511
20160619-C22 BC 8005-BC 7751
20160619-C23 BC 6099-BC 5997
20160619-C12 BC 5965-BC 5750
20160619-C24 BC 4181-BC 4040
F0
F1
F2
F3
F4
F5
e
Southwest wall
N40°W
20160610-C03 AD 135-AD 335
20160619-C26 AD 125-AD 315
20160610-C01 AD 575-AD 655
20160610-C02 AD 85 -AD 240
20160610-C07 BC 165-AD 20
20160610-C04 BC 90-AD 55
0701-C31 BC 860-BC 797
20160610-C05 BC 815-BC 780
0701-C32 BC 1696-BC 1611
20160610-C06 BC480-BC 390
20160610-C08 BC 3965-BC 3795
20160619-C17 BC 3815-BC 3709
20160619-C29 BC 7516-BC 7346
Large gravel
20160619-C16 BC 5393-BC 5320
0701-C33 BC 5366-BC 5282
0701-C34 BC 6231-BC 6084
20160610-C09 BC 4975-BC 4800
20160610-C11 BC4790-BC4685
20160610-C10 BC 5635-BC 5545
1m
Unit-1 Dark gray cultivated soil
Unit-2 Dark gray sandy soil
Unit-3 Gray sandy soil
Unit-4 Brownish gray sandy soil
Unit-5 Yellowish gray sandy soil with small gravel
Unit-6 Yellowish gray sand and gravel
Unit-7 Brownish gray sandy soil and gravel
Unit-8 Yellowish gray sandy soil and gravel
Unit-9 Yellowish gray sandy soil
Unit-10 Sand-gravel interbeded layer
Unit-11 Yellowish pumice
Unit-12 Brownish gray sand and gravel
Unit-13 Gray sand and gravel
Unit-14 Gray sandy soil with small gravel
C01~33 14C dating sample
F1 Fault

References

Headquarters for Earthquake Research Promotion (2016) Evaluation of Heisei 28 (2016) Kumamoto earthquake. http://www.mext.go.jp/component/a_menu/science/detail/__icsFiles/afieldfile/2016/04/15/1285728_001_1.pdf

Lin A, Chen P., Satsukawa T, Sado K, Takahashi N, Hirata S (2017) Millennium recurrence interval of morphogenic earthquakes on the seismogenic fault zone that triggered the 2016 Mw 7.1 Kumamoto earthquake, SW Japan. Bull Seismol Soc Am, in press. doi:https://doi.org/10.1785/0120170149

6 Destruction on Buildings and Houses Related to Coseismic Surface Rupturing

The Kumamoto earthquake series caused a maximum seismic intensity of 7 (on the Japanese 7-point seismic intensity scale) at both areas around the epicenters of the largest foreshock (M_w 6.1) and the M_w 7.1 main shock (Japan Meteorological Agency 2016), indicating severe damage and disruption throughout central Kyushu Island, including the Aso caldera region. These two large earthquakes caused more than 126,000 houses seriously damaged including 8300 destroyed (as of 1 July 2016) (National Research Institute for Earth Science and Disaster Resilience 2016). The damage and destruction of houses and infrastructures were considered to be caused by strong ground motions related to the coseismic surface ruptures and different geological environments. To share the information concerning the damage and destruction of houses and buildings, this book focuses on the damage of houses and buildings that are considered to be related with coseismic surface rupturing during the 2016 Kumamoto earthquake (Figs. 66, 67, 68, 69, 70, 71, 72, 73, 76, 77 and 78).

A. Lin, *The 2016 Mw 7.1 Kumamoto Earthquake*, https://doi.org/10.1007/978-981-10-5855-4_6

Fig. 66 Coseismic tension cracks developed throughout the house yard of the Volcanological Laboratory, Institute of Geothermal Sciences, Kyoto University (modified from Google Earth image acquired on 16 April 2016)

N
100 m
© 2016 ZENRIN

Fig. 67 Aerial photograph showing a close-up view of coseismic tension cracks shown in Fig. 66. Parallel–subparallel coseismic tension cracks developed in the house yard of the Volcanological Laboratory, Institute of Geothermal Sciences, Kyoto University (copyright permission by Asahi Shimbun)

Fig. 68 Houses and buildings damaged along coseismic cracks. (**a, b**) Eastward view of the damaged buildings of the Volcanological Laboratory, Institute of Geothermal Sciences, Kyoto University, where numerous of cracks and a large-scale landslide occurred on the slope. (**c**) Collapsed house on the preexisting fault scarp. (**d**) Earthquake-induced sand boil occurred along the coseismic crack extended to the base of a building

c

d

a

b

Fig. 69 Houses collapsed along a coseismic fault scarp. (**a**) Houses collapsed along a coseismic fault scarp of the Hinagu fault at the southern location of Takagi trench site. (**b–d**) Houses collapsed along the coseismic surface rupture zone developed along a preexisting fault scarp of the Futagawa Fault

c

d

a

b

Fig. 70 Houses collapsed along a coseismic surface rupture zone. (**a**) Houses collapsed along the coseismic surface ruptures along a preexisting fault scarp of the Futagawa fault. (**b–d**) Houses collapsed and slanted along the coseismic surface ruptures

c

d

a

b

Fig. 71 Houses damaged and collapsed along the coseismic surface ruptures. (**a, b**) House damaged on the coseismic surface ruptures. (**c**) Houses collapsed along the coseismic surface ruptures developed along a preexisting fault scarp. (**d**) House collapsed on a preexisting fault scarp

c

d

a

b

Fig. 72 Houses damaged on the coseismic surface ruptures. (**a–c**) Houses damaged on the coseismic surface ruptures. (**d**) Judgment card of dangerous house that cannot be inhabited again

c

d

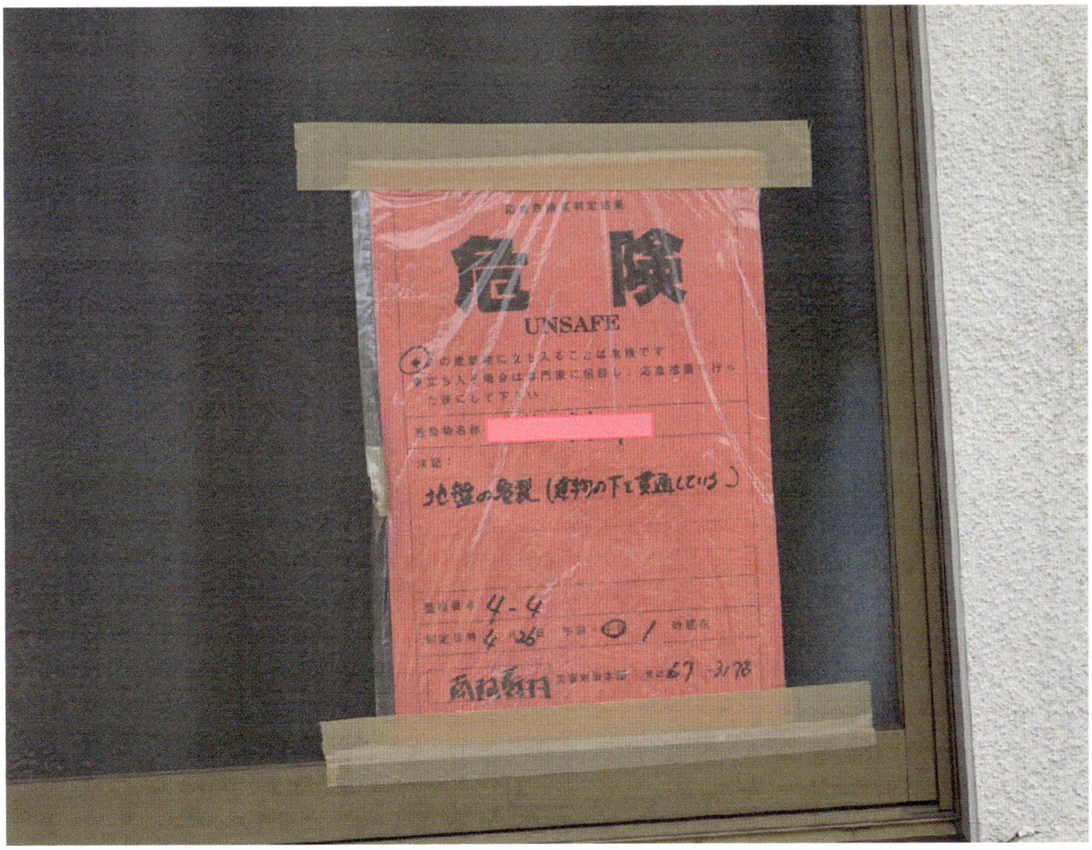

a

Fig. 73 Coseismic surface ruptures and damaged building of Tokai University at the Kumamoto Aso campus. (**a**) Coseismic surface ruptures occurred in a schoolyard throughout the base of the building. (**b**) Building damaged with coseismic cracks. (**c**) Coseismic surface ruptures occurred in a schoolyard

b

c

a

b

Fig. 74 Coseismic surface ruptures and damaged school house in a site near the Aso Nishi primary school. (**a**) Coseismic surface ruptures crosscut a road and fields (right side uplift ~50 cm). (**b**) A continuous coseismic fault scarp occurred in fields and crosscut a road (vertical offset ~1.7 m). (**c**) The building of the Aso Nishi primary school damaged on the coseismic surface rupture zone. (**d**) Close-up view of (**c**)

c

d

Fig. 75 Houses damaged along a coseismic graben. (**a**) Google image (acquired on 16 April 2016) showing a coseismic graben structure in which the houses and roads were damaged. (**b, c**) Ground subsidence inside a graben structure (the road was repaired immediately after the earthquake). The red roof houses shown in (**b–c**) are the same as that in (**a**), which are located inside a coseismic graben structure. (**d**) Damaged house on the hanging wall of a coseismic normal fault scarp located in the uplift side of graben

c

d

a

b

Fig. 76 Damaged houses on a coseismic normal fault scarp. (**a**) A damaged house located on a coseismic normal fault scarp. (**b**) Close-up view of (**a**). (**c**) A house inclined to a coseismic normal fault scarp, which was uplifted ~0.5 m by jacking. (**d**) House damaged on a coseismic normal fault scarp throughout a road

c

d

a

Fig. 77 Damaged building on a coseismic normal fault scarp. (**a**) A building on a coseismic normal fault scarp. (**b**) The building shown in (**a**) was uplifted ~0.5 ~ 0.7 m on a coseismic normal fault scarp (right side uplift). (**c**) Coseismic cracks occurred on a wall of building shown in (**a, b**)

b

c

a

b

Fig. 78 Damaged shrines and landslides. (**a**) Collapsed stone lanterns in the Hebi-ishi (*means* slake stone) shrine. (**b**) Collapsed Aso Shrine located in the northeastern end area of the coseismic surface rupture zone. (**c**) Landslides occurred in a slope of Aso caldera. (**d**) An overview of field investigation using drone to take photographs of coseismic surface ruptures

c

d

References

Japan Meteorological Agency (2016) Report on the Heisei 28 (2016) Kumamoto earthquake (No. 10). http://www.jma.go.jp/jma/menu/h28_kumamoto_jishin_menu.html

National Research Institute for Earth Science and Disaster Resilience (2016) 16 April 2016 Kumamoto Earthquake and its aftershock distribution. http://www.hinet.bosai.go.jp/topics/nw-kumamoto160416/?LANG=ja

Printed by Printforce, the Netherlands